THE INVISIBLE FARM

THE INVISIBLE FARM

THE WORLDWIDE DECLINE
OF FARM NEWS AND
AGRICULTURAL JOURNALISM
TRAINING

THOMAS F. PAWLICK

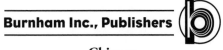

Burnham Inc., Publishers

Chicago

President: Kathleen Kusta
Vice-President: Brett J. Hallongren
General Manager: Richard O. Meade
Project Editor: Rachel Schick-Siegel
Designer: Tamra Phelps
Printer: Sheridan Books, Inc.
Cover Painting: "40 mph" by Lars-Birger Sponberg.
   Owned by Mr. and Mrs. Thomas Burns of Wixom, Mich.

Library of Congress Cataloging-in-Publication Data

Pawlick, Thomas.
      The invisible farm : the worldwide decline of farm news and
   agricultural journalism training / Thomas F. Pawlick.
         p. cm.
   Includes bibliographical references and index.
   ISBN 0-8304-1582-3 (paper: alk. paper)
         1. Journalism, Agricultural. 2. Journalism, Agricultural--Study and
   teaching. I. Title.

   PN4784.A3 P38 2001

                                                    2001025434

Manufactured in the United States of America

10 9 8 7 6 5 4 3 2 1

The paper used in this book meets the minimum
requirements of American National Standard for
Information Sciences—Permanence of Paper for
Printed Library Materials, ANSI Z39.48-1984.

For the crew of the good ship *Ceres*

Contents

Introduction   ix

1. The Invisible Farm   1

PART I

THE IMPORTANCE OF AGRICULTURE TO SOCIETY AND OF FARM NEWS TO THE GENERAL PUBLIC

9

2. Culture, Agriculture, and Survival   11

3. Rural Economies   23

4. Killing the Goose: Farming and Environment   41

5. Point of No Return: A Clash of World Views   75

PART II

THE IMPORTANCE ASSIGNED TO AGRICULTURE BY THE MAJOR NEWS MEDIA AND JOURNALISM EDUCATORS

81

6. Perceptions of Decline: Farm Journalism in
   North America   83

7. Chaos Unobserved: From *Kolkhozniki* to *Fermeri*
   in the East   105

8. Why Not Africa? Farm Journalism in the South   131

9. Conclusions   159

   Notes   163

   Selected Bibliography   183

   Index   195

Introduction

This book began life as a masters degree thesis in agricultural journalism, to be defended before a board of review at Carleton University in Ottawa, Canada—a board that included among its members a former Canadian federal minister of agriculture, Eugene Whelan. The intention was thus not to make waves, but to produce an orthodox, scholarly document that might prove useful to the academic community.

Reinforcing my initial desire to "stick to business" was the realization that, as a journalist covering agriculture for the Montreal *Gazette*, I had experienced previous run-ins with Mr. Whelan, some of them quite heated. The last thing on my mind was to revive the memory of past battles with one board member sitting in judgment on my thesis by creating new controversies.

Nevertheless, as the research went forward, it became more and more apparent that a suspicion that had been burrowing away in the recesses of my mind for several years was being confirmed in spades: Farm journalism was in fact a dying specialty. Coverage of agriculture and rural issues by the modern mass media had very nearly ceased.

Worse, this cessation of coverage had taken place at a point in the world's social and economic development where the whole nature of rural life and food production was being brutally overturned—altered so completely that once the process was done the possibility of return to past ways would be utterly foreclosed.

The stakes were high, not just for rural dwellers but for urban people as well. Everything—the environment, human health, social structures, the economy, politics, international trade—was affected. But nobody was noticing, least of all journalists.

And so the thesis became more than an academic document. It metamorphosed into a kind of alarm signal, a wake-up call to newsmen and women to see what was going on. Of course, most journalists today are urban-based and urban in outlook. For them, the rural world is not only *terra incognita*, but hardly worth reporting at all. Simply stating that a serious gap in coverage exists—without giving examples and explaining *why* it is serious—would be unlikely to convince members of such a notoriously skeptical profession. Newsmen and women, it seems, are all "from Missouri." They want to be shown.

Nor would one or two examples do. They might be exceptions. It had to be demonstrated that the importance of agriculture as a subject of news coverage is pervasive and universal: thus, the systematic marshalling of evidence in part I.

The detail provided, including sources cited from the agricultural literature, is specifically intended to give those journalists whose interest is piqued by the argument a substantial jumping-off point for research of their own—a base that those unfamiliar with the subject might waste considerable time trying to acquire without such leads. A primer, after all, should have some meat on it.

It might be objected that the survey of farm coverage and agricultural journalism training in North America, the former Soviet Bloc, and Africa, which is found in part II, leaves out areas of the world—Western Europe, Latin America, Asia, and Australasia—that are at least as important as those actually discussed. Undeniably, this is true. However, attempting to include every major global region would have resulted in a virtual encyclopedia of agriculture and agricultural journalism, running to thousands of pages. This was neither feasible nor desirable. Who would read such a tome? How many years would it take to write it?

I chose, instead, to concentrate on three regions that could be seen as approximately representative of the three major political and economic divisions in the contemporary world.

To his credit, and my great relief, ex-minister Whelan was not the kind of man to carry a grudge. As far as he was concerned, our previous public fights were water under the bridge. A farmer himself, he shared many of my views about the changing agricultural sector, and rural life, and was at least as worried as I was by the global implications. The thesis defense went well, and after it was over, he shook my hand. "I wish more media people were paying attention to this," he said.

So do I.

T. Pawlick
June 1, 2001

CHAPTER ONE

The Invisible Farm

In the late spring of 1973, writes Joyce Egginton in *The Poisoning of Michigan*, "a truck driver known as Shorty made a routine delivery from Michigan Chemical Corporation (MCC) to Farm Bureau Services, which operated the largest agricultural feed plant in Michigan."[1] The trailer load he dropped off was supposed to consist of 50-pound sacks of a feed additive called Nutrimaster, designed to aid cows' digestion and thus increase their milk output. But tragically, through an error at the MCC plant, the sacks were filled instead with a highly toxic fire retardant—a crystallized polybrominated biphenyl, or PBB.

The PBB was mixed into cattle feed, which was sold to farmers around the state, and within a few months an event comparable to the Love Canal or Chernobyl pollution disasters had occurred. Thousands of farm animals died—more than 35,000 contaminated cattle were slaughtered and buried at one mass burial site alone—and the environment near hundreds of affected farms was severely polluted. As for the human cost, virtually *every* resident of the state of Michigan—some 9 million people who drank milk or ate other farm products—was estimated to have absorbed "measurable levels" of PBB in their tissue. Hundreds developed immediate symptoms of severe toxicity for which there is no cure, while thousands of others were at risk of developing incurable diseases in the years to come.

According to Egginton, the response of the state's major news media to the disaster ranged from lukewarm to nonexistent. "The press, which should have been a public watch-dog, failed in its function," she charges. Her explanation is worth quoting at length:

Even more than politicians, newspaper writers are an urban breed. Large newspapers concentrate on city stories because therein lies circulation. Most journalists are out of their element on farms, and it was a long time before any Michigan editor was persuaded that there was a story worth chasing down dirt roads, not even marked on the state's highway map. . . .

Months passed before the state's two biggest newspapers, both in Detroit, tackled the PBB crisis in depth. The Detroit *Free Press* did three detailed articles, but not until March 1977—almost three years after the first quarantines. . . .

Out-of-state newspapers paid scant attention to the story, seldom printing sufficiently detailed accounts for their readers to fully understand the dimensions of the disaster. Americans who care about environmental causes were better informed about the dioxin contamination of Seveso, Italy, and the mercury poisoning in Minamata, Japan, than about the PBB crisis in Michigan, although this was the biggest chemical disaster and the worst man-made agricultural catastrophe in United States history. . . .

None of the national [television] networks tackled the subject as a documentary, although the idea was presented to them. Bonnie Pollard, senior associate editor of *Michigan Farmer*, made the suggestion in letters to all three networks late in 1974, but the correspondence was not even acknowledged . . . One argument used by television professionals was that sick cows are not good visual subjects, and since few viewers live on farms the topic would not attract enough interest.[2]

Only a tiny percentage of the 9 million people who were poisoned actually lived on farms, but the possibility that they might have been "interested" in knowing what had happened to them was apparently never entertained.

A Media Blind Spot

Unfortunately, the kind of urban-rural media blind spot of which Egginton complains is not rare. Since her book appeared, it has become increasingly the rule throughout much of the industrialized world, including Canada. According to the annual *Editor & Publisher International Yearbook*, for example, the number of Canadian daily newspapers listing an "agriculture editor" or "farm writer" among their full-time personnel dropped by 65 percent between 1975 and 1995. In the United States, even in a so-called "farm state" like Iowa, the number of dailies with full-time farm writers declined by 62 percent during the same period.[3] A comparison of the *Broadcasting Yearbook 1976* and *Broadcasting & Cable Yearbook 1994* shows that in the United States the number of on-air radio stations (AM and FM) listing themselves under the "agriculture and farm" format declined from an already low 3.4 percent in 1976 to a negligible 0.8 percent in 1994. In Canada in 1994, only three out of an existing national total 663 on-air AM and FM stations described their format as "agriculture and farm."[4]

The quality of agricultural news that does make it into print in North American newspapers and magazines is often low, as suggested in a 1994 study by the Office of Agricultural Communications and Education of the University of Illinois at Urbana-Champaign. The study's authors surveyed members of both the American Agricultural Editors' Association and the National Association of Agricultural Journalists, asking them to evaluate the quality of agricultural coverage in the daily press and the specialized farm press. The results:

> Both groups of judges said general-interest reporters' agricultural coverage is superficial, event-oriented, and often too cute or folksy. Both also said general reporters cover too few hard agricultural news stories, write too few in-depth stories, and do not give agriculture serious, long-term coverage. Furthermore, the judges said general reporters do not understand farming and give urban readers an incorrect picture of farming life.[5]

As for the specialized farm press:

> Both groups also agreed that farm magazines take a pro-industry point of view, run too many "successful farmer" stories and stories that serve advertiser interests, and fail to adequately investigate scandals . . . they agreed that magazines do not adequately cover environmental problems.[6]

In 1995, the Canadian branch of the media watchdog group Project Censored published its list of the "top 10 underreported stories," ongoing controversies that project staff believe have been largely ignored by the mainstream North American media. Three of the ten were farm stories.[7]

Reliable industry statistics are not available for the so-called "successor states" of the former Soviet Union (FSU), where a relatively high percentage of the population of many countries is still rural (well over 50 percent, for example, in the four Central Asian republics) or for most Third World countries, especially in sub-Saharan Africa, where farming is still the leading industry. Anecdotal evidence, however, indicates the state of coverage in these regions may be worse than that in North America (though perhaps for different reasons). A recent posting on the FSUmedia Internet list, for example, noted that:

> Russian farm journalists met outside Moscow to discuss the closure of rural papers and the disappearance of rural television and radio programs. According to delegates, only one national program devoted to farm issues remains. This news vacuum is particularly dangerous as parliamentary and presidential elections loom close. Approximately one-third of Russia's population is rural and has been traditionally politically active.[8]

Sociological factors arising from the history of Russian dominance in the region also tend to militate against adequate rural coverage. As John C. Merrill notes in *Global Journalism: Survey of International Communication*:

> In some of the countries, Russians dominated the cities while the local ethnic group lived in the countryside. This was true of Moldova, eastern Ukraine, northern Kazakhstan, and other places. What can happen is exemplified in Moldova . . . the Russian inhabitants speak only Russian, so read only the Russian press. The Russian-speaking urban population generally has more money and can more easily purchase the Russian-language press. It is also easier to distribute the press in the cities as compared to the more sparsely populated countryside.

Private investors generally come from the cities, too, and when they choose to invest in the media, they invest in the Russian media.[9]

Inevitably, events affecting the non-Russian-speaking rural population receive less coverage.

In sub-Saharan Africa, a rural press has only existed in most countries since the early 1970s, and most papers are struggling for survival. As few rural readers can afford to pay for subscriptions to magazines or newspapers, and advertisers are not attracted to publications whose readerships live at subsistence level, printed publications must depend on subsidies from national governments or external aid agencies. Many are no more than mimeographed newsletters, edited by poorly trained staff. As the authors of a UNESCO report, *Rural Journalism in Africa*, note: "Most of the rural newspapers are run by an editor assisted by one or two full-time reporters. Almost all are development officers in one or another agency, but none was ever formally trained to run or edit a newspaper."[10]

As for Africa's larger daily newspapers, a five-year survey in Nigeria found that less than 2 percent of the news hole (space for news as opposed to that for advertising) of the West African nation's major dailies was devoted to farm coverage—and that a great deal of what was published lacked "immediate functional relevance for the farming audience."[11] Concluded the survey author:

> It appears that Nigerian newspaper editors place more importance on revenue generating content than agricultural subject matter . . . editors may have assumed that farmers do not constitute an audience, and little can be done to persuade them to buy their "products." In other words, editors may regard agricultural content as "uneconomic" news.[12]

Whether in North America, the former Soviet Union, or Africa, the media resources devoted to coverage of agriculture and rural affairs are dwindling or inadequate. The reasons for this neglect differ from region to region, but the overall result is the same: major stories—not only environmental, but also economic, political, and socio-cultural stories—are being underreported. In some cases, they are not being reported at all.

Lack of Training

The tendency to neglect rural news is not only a function of the increasingly inadequate resources devoted by the major media to agriculture, of an editorial bias stemming from the dictates of a heavily-urban circulation base, or of the poverty of means in developing countries. It is also due to a lack of available training for journalists interested in covering the farm beat, as well as to the absence in general journalism education of efforts to alert students to the importance of agriculture to *all* readers— including those who live in the city.

For example, in all of Canada in 1995, there was only one post-secondary program in agricultural journalism, offered on a cooperative basis by Loyalist College of Applied Arts and Technology in Belleville, Ontario, and Kemptville College of Agricultural Technology, Kemptville, Ontario, both two-year community colleges. Due to funding cutbacks, the program was expected to be discontinued in 1996. In the United States, out of 510 university journalism faculties listed in the 1995 Association for Education in Journalism and Mass Communications (AEJMC) directory, only seven—.01 percent—offered courses in agricultural journalism. Three such schools were in the same state: Texas.[13]

As for the former East Bloc countries, a search of the literature, queries posted to the FSU (Former Soviet Union), EE (Eastern Europe), and Rusag (Russian Agriculture) Internet discussion lists, as well as numerous contacts with journalism educators active in the region in late 1995 failed to locate even a single journalism course directed specifically at training agricultural journalists. Typical was the response of Viera Simkova, of the Slovak Club of Agricultural Journalists: "There is no university in Slovakia that has training for agricultural journalists at this time. To be an agricultural journalist, not only in Slovakia but also in most other formerly communist countries is very difficult and unpopular now."[14]

Similarly, contacts with UNESCO, with the African Council on Communication Education (ACCE), and with such wide-based press organizations as Inter Press Service (IPS), though they located several university-level journalism programs and a few international development-oriented courses and workshops, revealed no program in sub-Saharan Africa directed specifically at training journalists to cover agriculture. The situation is only marginally better in other developing-country regions. According to IPS's Peter da Costa, the neglect of the subject is not due to lack of local interest, but to the reluctance of outside aid donors to allot funds for this purpose.

> Much as we would like to apply our . . . training specifically to agriculture, telematics and other global issues that are under-subscribed, we can't structure holistic, inclusive training programs based on our, or even on developmental priorities. Each donor [country or aid agency] has its priority, and its own funding gaps.[15]

Worldwide, most reporters and editors are urban-bred, urban-based, and urban-oriented, generations removed from farm life. As subsequent chapters will show, in the industrialized countries, this reflects the trend of the general population, while in the Third World it reflects the tendency of an educated urban elite to dominate the information industries. Lack of any personal experience with agriculture, or of the opportunity to learn about it in school, renders the countryside a literal *terra incognita* to many media people—out of sight and out of mind.

Even when an urban reporter or editor might be inclined to look at rural life, ignorance is a handicap. Compared to, say, the police or sports beats, about which any competent urban journalist with basic professional skills already has some familiarity, and whose finer points can be learned on-the-job, the complexity of the farm beat can take years to master: It covers an entire way of life, one often utterly foreign to city people.

The Invisible Farm

The paucity of resources made available by the major media to cover agriculture and rural affairs, and the ignorance of most journalists regarding rural issues, has rendered the farming and food distribution systems that feed the people of the globe effectively invisible. Massive and far-reaching changes now convulsing the so-called "harvest industries" around the world are proceeding largely without input from the general voting and consuming urban public—whose lives will be both directly and indirectly affected by the results. When these stories are reported, it is most often in scholarly or specialist publications, or the rare investigative expose, available to professionals who follow the industry but not to the population at large (many of the stories to be cited in subsequent chapters were documented from such sources).

Unaware of the issues, or of their own stake in what is happening, most newspaper readers, radio listeners, and television viewers know more about the private lives of Hollywood stars (few media organizations have farm editors, but virtually all have "entertainment" writers) than they do about the quality and stability of their own food supply. Nor are they aware of the ongoing, often universal, social and cultural changes—some subtle and some not—being provoked by the silent metamorphosis of our forgotten rural world.

As in the case of the poisoning of Michigan, the unwise assumption seems to have been made that no one is interested.

PART I

THE IMPORTANCE OF

AGRICULTURE TO SOCIETY

AND OF FARM NEWS

TO THE GENERAL PUBLIC

CHAPTER TWO

Culture, Agriculture, and Survival

The ultimate goal of farming is not the
growing of crops, but the cultivation and
perfection of human beings.[1]
—Masanobu Fukuoka

Among the first things one learns from the practice of agriculture, whether one comes to it by birthright—raised in the country—or transplanted from the city, is that farming is not a mere mechanical, scientific, or even economic enterprise, but a social and thus *cultural* one. The term agri*culture* implies this, as does the *Concise Oxford Dictionary*, whose 1964 edition defined "culture" as, first, "tillage; rearing, production (of bees, oysters, fish, bacteria)"—and only afterward as anything else.[2]

Perhaps the editors took their cue from anthropology, which accepts as given that the achievements of every civilized people since the dawn of the Holocene epoch 11,000 years ago, when farming began, have depended on agriculture as a base. As Barbara Bender says in *Farming and Prehistory*:

> Though definitions vary, all authorities would agree with Adams (1966, 38), who says that it is "a truism that complex, civilized societies depend upon a subsistence base that is sufficiently intensive and reliable to permit sedentary nucleated settlements, a circumstance that . . . in the long run has implied agriculture." We may add

11

that not only must there be farming, but in most cases it must be diversified and intensive.[3]

Later editions of the *Oxford* have revised their priorities in defining "culture," giving pride of place to "the arts [presumably including tillage, which is an art [4]] and other manifestations of human intellectual achievement," while moving "the cultivation of plants," "the rearing of bees, silkworms, etc.," and "the cultivation of the soil" to fourth place.[5] But the interconnections and overlaps between the realities of rural life as it is lived and the dictionary's various descriptions of culture remain as intricate as they are obvious. Even in the industrialized countries, almost every act a farmer performs—buying seed, repairing machinery, finding a market for a crop—has a socio/cultural component, governed by what the dictionary describes as "the customs, civilization and achievements of a particular time or people." When a baler breaks down at haying time, one customarily turns to a neighbor for help, or loses the crop. When a crop is lost, the banker (whose acts are ruled by cultural as well as financial norms) customarily forecloses. And, as I can attest from experience, nowhere more than in the country—where one is separated by greater physical distance from one's neighbors—does one feel so much part of a community.

But beyond the individual level, the level of farmer-with-neighbor, or the farmer-with-urban-interlocutor, agriculture has for millennia been intertwined with greater human society in myriad ways: as support and underpinning, as source, as inspiration—and most important culturally, as symbol.

Aside from the fact that all writers, artists, and musicians must eat, and thus depend at least indirectly on farmers, the place of farming and rural life in literature and the arts is inescapable. From Virgil's *Georgics* and Langland's *Piers Plowman*, through Shakespeare and Cervantes, to such varied twentieth century voices as France's Jean Giono, Britain's J.R.R. Tolkein, or Kenya's Ngugi wa Thiongo, the pastoral setting and rural theme have been integral to generations of written works around the world.

A bibliography drawn from English and North American literature alone would be as thick as a small telephone book, including, as it would have to, Milton, Cowper, Spenser, Herrick, Fielding, Goldsmith, Hardy, Cobbett, Coleridge, Lawrence, Wordsworth, (George) Eliot, Jefferies, Thoreau, Faulkner, Steinbeck, Frost, and

W. O. Mitchell—to name a few. Add to it a list from the other fine arts—the paintings of Turner, Constable, Millet, Van Gogh, the "pastoral" works of Beethoven (whose name means beet farm), Schubert, Sibelius, or Virgil Thompson ("The Plow that Broke the Plains")—and it would fill a thick phone book indeed. Folk and popular arts, from "bluegrass" music and the quilts and cake baking contests of annual farm fairs to the more sophisticated, stylized products of the Nashville "country music" scene in the United States, provide still more examples.

This artistic presence exists, and grows, because life in the country, on the farm, has assumed an Eden-like position in the mythologies and literatures of most human societies (John Steinbeck even uses the term in the title of his novel of California farm life, *East of Eden*).[6] As Raymond Williams points out in *The Country and the City*, "on the country has gathered the idea of a natural way of life: of peace, innocence and simple virtue."[7] Englishmen and Americans, in particular, have added their own mythology of the sturdy, independent yeoman farmer, the ultimate "freeman," beholden to no man, ready to defend his country and way of life, whether at Agincourt, where his longbow defeated the aristocratic French, or at Bunker Hill, where his flintlock rifle defeated the "Redcoats."

This, of course, is only half the mythological picture; the other half includes more hostile associations of the country with "backwardness, ignorance, limitation,"[8] and with the small town as incestuous snakepit (the work of Italian-Canadian author Nino Ricci, in *Lives of the Saints*, is a prime example)[9]. To many, rural means backward, and rural people are so many "Beverly Hillbillies" to be caricatured as hopeless bumpkins.

As the urbanization of North American society continues, even the rural milieu itself may become infected with the prejudice. Only recently, for example, the people of Calgary, Alberta—traditionally seen as the "cowtown" embodiment of Canada's western, cattle country culture—decided to end the city's yearly Cowboy Festival. Board members of the city's convention center, where the event was held, cancelled it. "There is a movement that would like to see Calgary in a more enhanced way—not just the cowtown, not just the oil and gas, but something more of a modern, competing world city," said board member John Schmal.[10]

But the positive myth is dominant.

Myth, of course, is nearly always based on a core reality. Yeomen really *did* defeat the French at Agincourt, and King George's military machine at Bunker Hill. The family farm really was a key engine of growth in the opening and development of North America by Europeans, and stamped its character on the democracies of the United States and Canada. Socially, politically, and economically it has been a pivotal entity in most societies around the world, from the "White Highlands" of Kenya's colonial period[11] to post-Second World War Japan, where land reform and economic empowerment of peasant farmers marked the first step in the change from semi-feudal state to modern democracy. Its disappearance could not help but have profound cultural consequences.

The cultural effects of radical change in agriculture could be more disturbing if, as Williams suggests, the significance of the mythical, pastoral model is not limited to the country, but applies also in the urban context:

> The complaints of rural change might come from threatened small proprietors, or from commoners, or even, in the 20th century, from a class of landlords, but it is fascinating to hear some of the same phrases—destruction of local community, the driving out of small men, indifference to settled and customary ways—in the innumerable campaigns about the effects of redevelopment, urban planning, airport and motorway systems, in so many 20th century towns and even, very strongly, in parts of London. I have heard a defence of Covent Garden, against plans for development, which repeated in almost every particular the defence of the commons in the period of parliamentary enclosures.[12]

The country is not revered only for itself, but as the original source and lasting symbol of an attitude toward life and one's fellow human beings. "People have often said 'the city' when they meant capitalism or bureaucracy or centralized power," writes Williams, and for its opposite, they use the code word "country."[13] Whether it is the destruction of a culturally integrating city neighborhood to make way for highway construction, or the elimination of a city's diverse restaurants in favor of internationally franchised fast-food competitors, the situation is essentially the same as it was when England's commons were enclosed. Williams' argument, which he expands to global proportions and in which the fall from Eden has such a strong allegorical echo, is worth quoting further:

> The very fact that the historical process, in some of its main features, is now effectively international, means that we have more than material for interesting comparisons. We are touching, and know that we are touching forms of a general crisis. Looking back, for example, on the English history, and especially its culmination in imperialism, I see in this process of the altering relations of country and city the driving force of a mode of production which has indeed transformed the world. I am then very willing to see the city as capitalism, as so many now do, if I can say also that this mode of production began, specifically, in the English rural economy and produced, there, many of the characteristic effects—increases of production, physical reordering of a totally available world, displacement of customary settlements, a human remnant and force which became a proletariat—which have since been seen, in many extending forms, in cities and colonies and the international system as a whole . . . What the oil companies do, what the mining companies do, is what landlords did, what plantation owners did and do . . . our images of country and city [are] ways of responding to a whole social development."[14]

Symbol under Seige

The idea that not only the rural world, but every world that could be considered humane or communitarian, is under seige—that we are continually being banished from a succession of socio-cultural Edens by the effects of the economic systems we ourselves construct—is shared by a variety of critics, including Murray Bookchin.

In his landmark 1976 essay, "Radical Agriculture," Bookchin insists that "in an epoch when food cultivation is reduced to mere industrial technique, it becomes especially important to dwell on the cultural implications of 'modern' agriculture—to indicate their impact not only on public health but also on humanity's relationship to nature and the relationship of human to human."[15]

It is not in dispute that agriculture is being reduced to an industrial technique, whose concentration, mechanization, and capitalization have so "rationalized labor" as to wipe out the rural populations of most of North America, large parts of Europe, and Japan. The statistics are inescapable. For example, a decades-long trend of rural out-migration in Canada became particularly intense after the Second World War. Between 1950 and 1980, the number of people living on farms "was slashed by a full 50 percent. In Ontario alone, nearly 362,000 people left the land—the equivalent in urban terms of the entire population of a city the size of Hamilton—suburbs included—packing up and walking away from their homes."[16]

According to United Nations Food and Agriculture Organization (FAO) figures, the population of the United States (the majority of whose people were engaged in agriculture at the time of Henry David Thoreau) included only 2 percent engaged in farming in 1993.[17] Similar movements have occurred in France, Germany, and Japan, with significant drops in agricultural population between 1961 and 1993. Overall, the European Union (EU) has seen its agricultural population—already considerably reduced from pre-Second World War levels—drop by 14 percent between 1961 and 1993.

According to the FAO, the agricultural population of the Soviet Union, including Russia, also fell steadily from 1963 to 1984, as farming became increasingly mechanized and the size of state and collective farms grew. It has kept falling in most successor states since the dissolution of the Soviet empire. In many Third World countries, only massive growth in the general population has prevented a similar drop in rural numbers. Migration from the impoverished countryside to cities proceeds at an unsustainable rate, causing acute urban environmental and employment problems,[18] but the rural population continues to grow faster than its members can flee to the towns. In Kenya, total population grew from 8,592,000 in 1961 to 26,090,000 in 1993—an astonishing increase of 303 percent—while agricultural population jumped from 7,473,000 to 19,737,000 in the same period—a 264 percent rise.[19] The country's rural economy, crippled by falling world prices for such commodities as coffee, tea, and sugar and the unfavorable terms of trade created by the recent Uruguay Round negotiations under the General Agreement on Tariffs and Trade (GATT), cannot provide for its newest arrivals—nor can the overstrained urban environments of Nairobi and Mombasa to which so many migrate.

Throughout the twentieth century, all over the world, agriculture has shed labor in massive numbers, first depopulating the countrysides of most northern nations, and now beginning to drain those of the global South and to complete the process of concentration—already well-launched by communism's state and collective farm systems—in the poorer parts of the former East Bloc. The small family farm has been replaced by larger and larger private operations, and finally by the corporate farm. As *Washington Post* reporter Nick Kotz writes, observing the U.S. scene:

> The medium to large-size "family farms"—annual sales of $20,000 to $500,000—survived earlier industrial and scientific revolutions in agriculture. They now face a financial revolution in which the traditional functions of the food supply system are being reshuffled, combined, and coordinated by corporate giants. "Farming is moving with full speed toward becoming part of an integrated market-production system," says Eric Thor, an outspoken farm economist. . . . "This system, once it is developed, will be the same as industrialized systems in other U.S. industries". . . . Twenty large corporations now control [all of U.S.] poultry production.[20]

Describing the entry of oil and chemical companies, including the giant conglomerate Tenneco Inc., into farming, Kotz asks: "Will agriculture become—like steel, autos and chemicals—an industry dominated by giant conglomerate corporations like Tenneco? In that case the nation will have lost its prized Jeffersonian ideal, praised in myth and song, of the yeoman farmer and independent landowner as the backbone of America."[21] The industrialization of agriculture, he writes, has further serious implications:

1. The future shape of the American landscape. Already in this country, 74 percent of the population lives on only one percent of the land. If present trends continue, only 12 percent of the American people will live in communities of less than 100,000 by the 21st century; 60 percent will be living in four huge megalopoli, and 28 percent will be in other large cities;

17

2. The further erosion of rural life, already seriously undermined by urban migration. Today 800,000 people a year are migrating from the countryside to the cities. Between 1960 and 1970 more than half our rural counties suffered population declines. One result is the aggravation of urban pathology—congestion, pollution, welfare problems, crime, the whole catalog of city ills;

3. The domination of what is left of rural America by agribusiness corporations. This is not only increasing the amount of productive land in the hands of the few, but is also accelerating the migration patterns of recent decades and raising the specter of a kind of 20th-century agricultural feudalism in the culture that remains.[22]

Of course, the same trends Kotz deplores are working—or already have worked—the same kinds of changes in the industrialized economies of Western Europe and Japan. Bookchin laments the increasingly common worldwide result:

Agriculture, in effect, differs no more from any branch of industry than does steelmaking or automobile production. . . . In this impersonal domain of food production, it is not surprising to find a "farmer" often turns out to be an airplane pilot who dusts crops with pesticides, a chemist who treats soil as a lifeless repository for inorganic compounds, an operator of immense agricultural machines who is more familiar with engines than botany, and, perhaps most decisively, a financier whose knowledge of land may beggar that of an urban cab driver. Food, in turn, reaches the consumer in containers and in forms so modified and denatured as to bear scant resemblance to the original. In the modern, glistening supermarket, the buyer walks dreamily through a spectacle of packaged materials in which the pictures of plants, meat and dairy foods replace the life-forms from which they are derived. The fetish assumes the form of the real phenomenon. Here, the individual's relationship to one of the most intimate of natural experiences—the nutriments indispensable to life—is divorced from its roots in the totality of nature. . . . This denatured outlook stands

18

sharply at odds with an earlier animistic sensibility that viewed land as an inalienable, almost sacred domain, food cultivation as a spiritual activity, and food consumption as a hallowed social ritual.[23]

American poet and social critic Wendell Berry, himself a farmer, has also identified and warned against this process of agricultural industrialization and rural depopulation, which he calls "a work of monstrous ignorance and irresponsibility on the part of the experts and politicians, who have prescribed, encouraged and applauded the disintegration of farming communities all over the country."[24] Like Kotz and Bookchin, Berry sees links between the rural and urban crises, and fears the cultural effects:

> Few people whose testimony would have mattered have seen the connection between the "modernization" of agricultural techniques and the disintegration of the culture and the communities of farming—and the consequent disintegration of the structures of urban life. What we have called agricultural progress has, in fact, involved the forcible displacement of millions of people.
>
> I remember, during the 50s, the outrage with which our political leaders spoke of the forced removal of the populations of villages in communist countries. I also remember that at the same time, in Washington, the word on farming was "Get big or get out"—a policy which is still in effect and which has taken an enormous toll. The only difference is that of method: the force used by the communists was military; with us, it has been economic—a "free market" in which the freest were the richest. The attitudes are equally cruel, and I believe that the results will prove equally damaging, not just to the concerns and values of the human spirit, but to the practicalities of survival. . . . The aim of bigness implies not one aim that is not socially and culturally destructive.[25]

Berry insists food is "a cultural product; it cannot be produced by technology alone"—that is, not unless the process is radically simplified, as it is in highly mechanized, industrial monocropping (single-crop) systems. Massive acreages are leveled and sown, year after year, with no or only infrequent crop rotations, to a lone, high-cash-return crop such as hybrid maize, which quickly depletes soil nutrients. To make up for the lost nutrients, especially nitrogen, heavy doses of inorganic chemical fertilizers are employed, which "burn" living soil organisms and pollute the water table. The industrial division of labor involved in such environmentally destructive "factory farming" also multiplies the number of wage-worker "specialists" doing the work, each focused on his or her narrowly defined task, while eliminating generalists who, like the vanishing family farmer, can envision whole systems. The process—though few of the right-wing economists who support it are likely to admit the fact— is remarkably similar in its end result to the collectivization of agriculture in the former Soviet Union (see chapter 7), and the destruction of the so-called Kulaks.

An article in the June 23, 2000, issue of *The Windsor Star* gave an example of what the drive towards bigness-for-the-sake-of-bigness can lead to:

> Picture the cultivated fields of four large traditional farms under a canopy of Dutch glass.
>
> Such is the scope of a massive, 107-acre greenhouse project Leamington developer John Stockwell proposes to build in Lakeshore Township between Highway 401 and County Road 42, making the Lakeshore development the largest in the region. Once complete, it would be 20 percent larger than all the land at Devonshire Mall, including parking lots.
>
> The year-round, continuous production veggie factory, developed by Agro-Sun Technologies, would employ 435 workers and provide an $18 million payroll. Salaries would range from $27,586 for hourly employees to $66,600 for managers.
>
> In an interview, developer John Stockwell said the scale of the $170 million project is so large that a motel-style housing development is included to accommodate 350 full-time employees.[26]

Like Williams, Wendell Berry sees this process as symbolic of a social outlook that now runs through virtually every aspect of Western life, one that favors compartmentalization and leads to "a radical simplification of mind and character":

That the discipline of agriculture should have been so divorced from other disciplines has its immediate cause in the compartmental structure of the universities, in which complementary, mutually sustaining and enriching disciplines are divided, according to "professions," into fragmented, one-eyed specialties. It is suggested . . . that farming shall be the responsibility only of the college of agriculture, that law shall be in the sole charge of the professors of law, that morality shall be taken care of by the philosophy department, reading by the English department, and so on. The same, of course, is true of government, which has become another way of institutionalizing the same fragmentation. . . . However, if we conceive of culture as one body, which it is, we see that all of its disciplines are everybody's business.[27]

The "compartmental" mind-set, symbolized by the factory-farm, is symptomatic of a culture of alienation, whose components are cut off from one another as well as from nature. But, Berry asserts, "a culture cannot survive long at the expense of either its agricultural or its natural sources. To live at the expense of the source of life is obviously suicidal."[28] By way of example, he points to the comments of former U.S. Secretary of Agriculture Earl Butz, and former Secretary of Defense Robert McNamara:

Our recent secretary of agriculture remarked that "Food is a weapon." This was given a fearful symmetry indeed when, in discussing the possible use of nuclear weapons, a secretary of defense spoke of "palatable" levels of devastation. Consider the associations that have since ancient times clustered around the idea of food—associations of mutual care, generosity, neighborliness, festivity, communal joy, religious ceremony—and you will see that these two secretaries represent a cultural catastrophe. The concerns of farming and those of war, once thought to be diametrically opposed, have become identical.[29]

If what such critics claim is true, a kind of circle has been closed. Farming, the basis of settled human life, has permitted the development of a civilization that alienates and cuts human beings off from life. By helping depopulate the land, to overcrowd and

21

overwhelm the cities and degrade the natural environment, the farming systems developed by our culture run the risk of destroying themselves—and thus the foundation of our civilization.

Perhaps symptomatic of what is happening in terms of social disintegration is the report, in July 2000, that crack cocaine use is "on the rise" in rural Canada.

> It's not just an inner-city drug anymore. Addictions experts from across the country say crack cocaine is fast becoming a fixture in rural and small town Canada, and its users—many of them addicts—are getting younger.
>
> Crack, a cheaper form of cocaine usually associated with the back alleys of major cities, has spread into small towns and cities in recent years, the experts say, and its buyers are increasingly teenagers and people in their twenties.
>
> "The spread is quite dramatic," said Doug Smith, program coordinator of Toronto East General's detox unit.[30]

"This is not merely history," writes Berry. "It is a parable."[31] And where ought our society's parables be told, if not in the press?

CHAPTER THREE

Rural Economies

*Agencies should listen to and learn from
the rural poor, not from blinkered economists
trapped in their own ideologies.*[1]
—Riad El-Ghonemy

The global debt crisis of the 1980s—during which international prices of agricultural commodities fell to their lowest levels in 50 years,[2] and the World Bank/International Monetary Fund (IMF)-imposed structural adjustment programs that resulted, had tremendous impacts on farming and rural populations around the world.

But the Uruguay Round of negotiations under the General Agreement on Tariffs and Trade (GATT), and the new World Trade Organization (WTO) that was established to regulate the agreement, are likely to have even more far-reaching effects. These will be accentuated by similar, regional trade pacts like the North American Free Trade Agreement (NAFTA) among Canada, the United States, and Mexico.

After nearly a half-century of domestic protection, agriculture is being thrust suddenly—some think brutally—into the international free-trade arena.

Article XI of the original post-Second World War GATT agreement, signed in 1947, specifically excluded agricultural products from the list of international trade items on which it was forbidden to levy import quotas. This exemption "loophole" was enlarged in 1955—at the insistence of the United States—under the so-called Section 22 waiver, supposedly granted "temporarily" by GATT negotiators, but continued for 30 years afterward.[3] As a result of the exemptions, nations signatory to the GATT could protect strategically, culturally, or politically important domestic agricultural industries and rural populations from foreign competition—fair or unfair—that might threaten them. As a UN Food and Agriculture Organization (FAO) study explains:

> Agricultural trade policy has long reflected the widely held belief that, because of its importance and vulnerability, the agricultural sector could not be exposed to the full rigors of international competition without incurring unacceptable political, social and economic consequences.[4]

No more. For the first time in postwar history, not only was agriculture a major element in trade negotiations, but also the final GATT Agricultural Agreement brought the industry firmly under the so-called "new world order" in international trade. Implementation of the Agricultural Agreement will be overseen by a Committee on Agriculture, whose powers will be spelled out by the WTO. Depending on one's viewpoint, the effects could prove either catastrophic or beneficial—but few observers doubt that over the long run they are certain to be sweeping.

If the people of the countries affected are to assess clearly whether the GATT's results are something they really want, it is imperative that their media report the changes adequately. Not only farmers and traders, but also everyone who consumes food has a stake in the outcome.

A New Regime

Research data and studies produced by the FAO formed a key part of the material considered by trade representatives at the GATT talks, and FAO experts observed the bargaining at firsthand. As the organization reported:

> The negotiations turned out to be a contest between the advocates of fundamental reform and those of trade lib-

24

eralization. . . . Advocates of reform were striving to get the GATT rules revised to remove the special exceptions for agriculture that accommodated the domestic farm policies of the major trading powers. Indeed, the reformers wanted GATT rules for agriculture to go further than rules for non-agricultural industries by barring trade-distorting domestic subsidies as well as special border protection and export subsidies.

The advocates of trade liberalization . . . argued that countries should have a right to pursue national agricultural policies that suited their own particular agricultural conditions, but that these policies should be gradually modified to limit or reduce their adverse impacts on trade. . . .

In the end, there were more reforms than liberalization in the Uruguay Round.[5]

Of course, appropriation by the negotiators of such euphemistic terms as "reform" (which implies improvement) and "liberalization" (which implies openness and magnanimity) tended to beg a number of questions. There was no guarantee that "reform" would improve things for farmers or consumers, while the "openness" of retaining border controls was disputable.

Several elements of the final agreement's import protection regime affected farming, but the most important dealt with tariffication, a process that converts all presumed import barriers—even indirect ones—into quantifiable tariffs. The FAO report summarizes:

The new rules required that all quotas, variable levies and other import barriers be converted to common tariffs, as soon as the agreement took effect. These and existing tariffs had then to be reduced by a minimum of 15 percent each over the implementation period with the tariff reductions as a whole having to average 36 percent. Developing countries were required to reduce tariffs by 24 percent and were allowed 10 years, instead of six, to implement the cuts. . . . The tariffication of non-tariff barriers and the prohibition against future use of such non-tariff instruments represents a major reform of the trade rules affecting agriculture. It should bring transparency to barriers that have been hidden from public view and should also expose the high levels of protection enjoyed by agricultural producers in some countries.[6]

Most commentary on the agreement, particularly by spokespeople representing the conservative, corporate agribusiness view, applauded the new, open-borders regime. For example, economist Stefan Tangermann of the University of Gottingen, Germany, was quoted widely to the effect that the Agreement on Agriculture was "a historical breakthrough . . . a major step in a good direction."[7] He was especially effusive about the provisions on tariffication:

> The substitution of bound tariffs for the wide variety of protective border measures which governments could use at will to shield their farmers against international competition creates a new degree of predictability in agricultural trade. For example, in the beef market the European Union will no longer be able to set threshhold prices, and charge variable levies, at levels determined solely by domestic considerations. . . . Moreover, a complete closure of markets is no longer possible, as countries have to provide minimum access opportunities.[8]

Nor would the barriers at international borders be the only ones to come tumbling. Calling attention to a feature overlooked by many other observers, Tangermann added:

> Where protection against imports can no longer be provided at will and export subsidies will have to be cut back, there is the danger that domestic subsidies will be used instead. It is therefore good to know that there are now also limits to the extent to which governments can provide domestic support.[9]

The assumption of Tangermann, and spokespersons for most G-7 countries (France was a notable exception) whose trade representatives dominated the GATT bargaining, was that removing protection for domestic farm industries would—automatically—improve the previous system.

Whose Improvement?

The question "improvement for whom?" went unspoken, except by a minority of critics representing environmental or consumer groups and the welfare of family, as opposed to corporate/industrial, farm interests. These included British economist/authors Timothy Lang and Colin Hines, whose book, *The New Protectionism*, attacks free-trade assumptions.[10] Lang and Hines were blunt in a January 1996 article that followed publication of their book:

> The new GATT has enshrined the 1970s Reagan/Thatcher era formulae: deregulation, economic efficiency, international competitiveness, and a dogmatic reliance on the market to meet all needs. Yet there is much evidence these policies don't work for the common good. The large farmer, traders and big companies benefit, but the evidence is that intensive, high-input farming, the logical outcome of these policies, is disastrous for the environment, rural economies, food quality and food security.[11]

The authors' references to the Reagan-Thatcher era and rural economies went to the heart of the problem. As Terry Pugh, former editor of the Canadian National Farmers Union publication *Union Farmer*, argues, the macroeconomic views of the GATT negotiators, like those of the American side in the bargaining that preceded NAFTA, more or less faithfully represented the so-called "neo-classical" or "neo-liberal" approach to economics, inspired by such economists as Milton Friedman and the Chicago School of Economics. This approach has become dominant in the past two decades.[12] For example, most of the administrative and financial officers appointed to key positions at the World Bank and IMF during the Reagan/Thatcher period—who now dominate policy-making at both institutions, as well as the top economic policy echelons of the United States and several other G-7 governments—are sympathetic to the Chicago School,[13] which claims among its intellectual antecedents nineteenth-century "classical" or "liberal" economists Adam Smith and David Ricardo, as well as the philosopher David Hume and Utilitarian social theorists James Mill and John Stuart Mill.

The credibility of the classical economic worldview—which regarded government regulation or intervention of any kind as, at best, a necessary evil and advocated free trade and a *laissez-faire* approach to business and society—was severely battered during the Great Depression of the 1930s. The depth of suffering caused by that worldwide economic disaster made it almost impossible to defend the idea that economies were automatically self-regulating and required no government intervention to arrive at the Mills' much-quoted "greatest good for the greatest number." By the late 1930s and early 1940s, the theories of Smith and Ricardo were largely superseded by the interventionist economic thinking of England's John Maynard Keynes, whose theories not only formed the backdrop for the Roosevelt-era New Deal, but for most government interventions in the world's industrial economies for a generation afterward.

The election of Reagan in the United States and Thatcher in the United Kingdom reversed the situation. Their appointees to key national and international posts effectively turned back the economic clock, rejecting Keynes' theories of "pump-priming" and deficit spending, and resurrecting the system of ideas that had been prevalent prior to the Great Depression—ideas that tend to be shared by the administrators of many of the largest transnational corporations dealing in farm products.

Those ideas, Pugh believes, dominated the thinking of GATT negotiators:

> It is no coincidence that the final GATT pact embraced these (free-trade, market-oriented) principles, since the neo-classical, free market ideology permeating the report was shared by negotiators from the industrialized nations, particularly the U.S. In fact, the formal and informal ties between corporate executives and high-level government bureaucrats at the GATT talks was uncharacteristically illustrated very openly in December 1993, when the U.S. government sent an "advisory group" of 11 corporate presidents and chief executive officers to "monitor" the progress of the talks and provide "valuable guidance" on what the agreement should contain with regard to agriculture. Foremost among the advisory group was Whitney MacMillan of Cargill Grain.[14]

Cargill Grain's prominence at the GATT talks spoke volumes to knowledgeable observers, particularly those familiar with the marketing board system of farm product sales employed in Canada and some European and developing countries. The new GATT directly threatens the board system, while favoring the transnational corporate sector. Boards are even more directly threatened by NAFTA provisions, as Canadians discovered in 1995–1996 when the United States filed a brief under NAFTA terms demanding that Canada quash border tariffs designed to protect farmer-members of Canada's dairy, egg, chicken, and turkey marketing agencies.[15]

Cost-Price Squeeze

Marketing boards were created in countries like Canada in direct response to an oligopoly situation in which a small number of large companies controlled the food sector, putting individual farmers in a "cost-price squeeze." As the author of this text noted in an earlier report:

> Today's average [food industry] corporation is not only multinational and multi-industry (horizontal integration), but involved within each industry in the entire production chain from basic supplies to product marketing (vertically integrated). Needless to say, the individual farm operator—who according to Revenue Canada in 1979 boasted an average net income of only $12,598— has become an almost insignificant economic cipher, overwhelmed by the financial might of those with whom he must deal.
>
> And yet he has no choice but to deal. On the one hand he must buy equipment and supplies—machinery, fuel, feed, seed and fertilizer—which along with rocketing land prices and the interest on his loans make up the farmer's cost factor. On the other hand, once a crop is raised it must be sold to buyers in the processing, distribution and retail (PDR) sector who influence the farm gate price the producer receives—the second factor in the squeeze. Obviously, the contest is unequal at both ends of the production cycle.[16]

Large corporations set prices and control the delivery of the famer's inputs, from machinery to fertilizer, and would dictate the price he is paid for his produce—if it wasn't for the boards' mediating function. Marketing boards group individual producers of a given farm commodity, such as eggs or wheat, who pool their production and agree to sell at the same price. This provides a countervailing weight to that of the corporations and helps the board's farmer-members to ease the cost-price squeeze that would otherwise put many of them out of business. In effect:

> Many boards serve, in part, as farmers' unions. Federal board officials are usually government-appointed, but most provincial and local boards are elected by their farmer-members, just as the officers of a union local would be. They negotiate, sometimes demand, a return on farmers' labor in much the same way that a union bargains for workers' wages.[17]

This bargaining ability, of course, runs counter to the interests of the corporations with which boards deal. Economic self-interest should impel the corporations to try to neutralize or eliminate marketing boards. Pugh summarizes the GATT talks' underlying agenda succinctly, making the significance of the presence of Cargill's Whitney MacMillan at the trade talks self-evident:

> In the grain sector, for example, a half-dozen corporations influence global prices and supplies, and design trade policies which accommodate their self-interest. The largest of this oligopoly is Cargill Grain, closely followed by Archer Daniels Midland (ADM), Continental, Louis Dreyfus, Bunge & Borne, Mitsui and Feruzzi. These companies' overriding need to source raw materials from the cheapest suppliers, and acquire unfettered access to expanding markets, provided the initial impetus and ongoing direction for GATT negotiations. Domestic farm programs, including Canada's orderly marketing and supply-management systems, represent obstacles to this flow of commodities and capital. Consequently, marketing boards were targeted for elimination in the trade talks.[18]

30

In the end, GATT provisions paid lip-service to the idea of domestic marketing boards, but severely undercut their effectiveness and—observers like Lang, Hines, and Pugh argue—laid the groundwork for the boards' eventual destruction. The concept that provided an opening was that of "indirect barriers to trade." According to U.S. negotiators (who applied the same logic in negotiations leading to the NAFTA treaty), any marketing or regulating system that impacted in any way on international trade could be classed as an "indirect barrier." This included domestic food safety regulations, as well as such domestic marketing arrangements as farm product marketing boards.

In theory, a domestic food safety rule more stringent than commonly accepted international regulations could be called a barrier to trade if it forbids imports of foods or food products deemed unsafe for consumption in one country, but not elsewhere. Regulation of supply or setting of prices in a domestic market by a domestic board could also be a trade barrier, if the products are also sold abroad.

According to the GATT, such arrangements, if their status as barriers can be established, must be converted to cash tariffs—and the tariffs reduced in the increments quoted in the FAO report (p. 32). Theoretically, these tariffs could in the future be cut still further, to negligible amounts—even zero—rendering national food safety rules and domestic marketing boards effectively useless. NAFTA provisions demand even more severe reductions than the GATT. Domestic boards, Pugh wrote:

> are non-trade-distorting because they don't sell internationally. In a sense, they act as collective bargaining agents on behalf of farmers in negotiations with Canadian-based food processors, and set prices according to a formula which takes account of the cost of production of the most efficient producers. This encourages national commodity self-sufficiency, while assuring consumers a steady supply of quality food at reasonable prices. . . .
>
> Although Canada's supply-management boards do not distort global trade, they nevertheless block the expansion of transnational corporate traders and processors—and are thus classified as non-tariff barriers to

31

trade. Canadian-owned agribusiness corporations stand to gain tremendously from their demise. It is the family farm, rural communities and the environment that will suffer.[19]

Operators of small- to medium-sized family farms will suffer by losing their only defense—the bargaining leverage of their boards—against the heavyweights of international trade, and by being faced with a consequently worsened cost/price squeeze. World prices for farm commodities, forced down by the combined influence of multinational trading corporations, will eventually become the domestic norm, slashing individual farmers' returns to levels below the cost of production. This will force increasing numbers of them out of business, worsening the depopulation of the countryside mentioned earlier (pp. 20–22), and forcing the collapse of the economic base of many rural communities.

A Micro-Level Example

A graphic example of the process, sparked in this case by the NAFTA pact rather than GATT, was given in a Toronto *Globe and Mail* article describing the shutdown of a Kakabeka, Ontario, dairy farm. One of very few reports in mainstream Canadian media to discuss the subject on a microeconomic level,[20] it is worth quoting at length:

> "Well," said my neighbor, settling heavily into her chair, "that's it; we've sold the cows."
>
> The rest of the ladies' Scrabble group took a minute to absorb the news. Sold the cows? No more dairy farming? . . .
>
> The news was a shock to us all. It was one of those times when we didn't know what to say. They'd been doing well at it, well enough to support her and her husband, their four sons, his parents and his younger brother. Two houses, two barns, two farms, four silos, 50 cows, a flock of chickens, hay fields rented from neighbors—it was quite an operation. Of the five dairy farms left in our township, I'd have said theirs was the best. . . .

The farm first belonged to his parents. When they came here from Holland in the early 50s, they took over a farm that had been neglected and made it into a wonderful example of an efficient dairy operation. It was something to be proud of. The farm is one of the oldest in the township, having been started in about 1895, when the first settlers were trying to make a go of it in this inhospitable region of Ontario.[21]

Why should such an efficient farm fail? The answer is the relentless cost/price squeeze, and its tendency to force even the best farmers to "get big or get out," as former U.S. Secretary of Agriculture Butz advised. Constant expansion in an effort to achieve economies of scale—buying out neighboring farms, purchasing larger machinery, hiring seasonal labor—is often the only choice in face of such economic pressures. The article continues:

"We're young enough to do something else," my neighbor explained. "If we want to stay in farming, we'd have to expand so much that we'd be in debt for 20 years. We'd be lucky to have it paid off by the time we retire. We'd be working as hard as we work now but it would all go to paying off debt. We'd need to have more cows, more hay to feed the extra cows, more expenses for new regulations—it's too much". . . .

I think she's still in shock. I know I am. It's sad to see the end of something, and maybe sadder still to know that if they can't make a go of it, neither will the four farms still operating here. Those four are just postponing the inevitable.

Next year, changes resulting from the free trade agreement will mean that Canadian dairy farmers will no longer enjoy the protection of any sort of regulation. We will see more of our milk and dairy products coming from the U.S., and eventually, when I am in a gloomy mood, all of North America will be served by three dairy farms, each the size of Nova Scotia.[22]

The article author's fears may be less exaggerated than she realizes. According to another *Globe and Mail* article, published on page one on April 30, 1996, if the United States wins its NAFTA case against Canada, the existence of some 32,000 farms could be endangered.[23] (It is revealing to note that the *Globe and Mail* may be the only major Canadian daily to have given consistent coverage to the NAFTA farm story. For example, a review of the April 30 and May 1, 1996 issues of the Vancouver *Sun*, *Financial Post*, Ottawa *Citizen*, Montreal *La Presse*, Montreal *Gazette*, and Kingston *Whig-Standard* located no reference to the trade dispute, despite its potentially disastrous effects on Canadian farmers.)

In the United States, where the idea of marketing boards is considered vaguely "pinko" or "commie" (as one Michigan farmer commented to this author), corporate domination of the farm sector is a fact of life. The few independent hog farmers left in Missouri have been hurt severely by the new terms of trade in the global system, as a *USA Today* article published in the spring of 2000 reported:

> Family farms, which have been declining in number for decades, continue to struggle as the agricultural industry's revolution roars on. Farms continue to get bigger as corporations make them more efficient, and the independent growers are groping for ways to increase production, computerize their operations and compete globally. Many are losing the battle. . . .
>
> One reason: the introduction of corporate operations such as Premium Standard Farms and Smithfield have changed the economics of the industry. . . . "Global economics is part of the driver here," said Thomas Hoening, president of the Federal Reserve Bank of Kansas City.[24]

The article noted that Premium Standard's Missouri hog farming operation occupies no fewer than *55,000 acres*, compared to the mere 2,000-acre farm of independent operator Wes Shoemyer. "The industry doesn't reward small farmers anymore, so people are getting out," Shoemyer said.

Social Catastrophe

In fact, in North America and Western Europe the farm depopulation process has already come close to running its historic course. As

Williams[25] and Wallerstein[26] point out, the rural-urban migratory trend began in Europe centuries ago, and was catalyzed by the Industrial Revolution. Rural labor shed over the years has been absorbed in the northern industrialized countries, with varying degrees of success or dislocation (sometimes horrendous dislocation), first by urban manufacturing and most recently by service industries.

However, no such urban manufacturing and service industry infrastructure exists in most of the Third World, or in parts of the former East Bloc where a similar rural-urban exodus is underway. Nor is one likely to evolve quickly under a free-trade regime that favors the "comparative advantage" of the already well-established industries of the wealthy G-7 countries. As Lang and Hines warn, the results could be socially catastrophic:

> If the North's development model is imposed and fosters the same rural/urban spread of population in the South already seen in some of the richest, "efficient" agricultural systems like Canada, Australia or the UK, 1.9 billion of the planet's rural dwellers will end up living in towns. To do what? Fed by whom?
>
> Evidence, North and South, suggests these new urban dwellers, forced out of agriculture by the new free trade patterns, might become cheap urban labor if they're lucky—but more likely just hungry, unemployed survivors. One need not be a neo-Malthusian to view this scenario with trepidation.[27]

The social squalor portrayed in Dickens' *Hard Times* and *Oliver Twist*, based on England's experience with the effects of the Industrial Revolution, pale in comparison.

As for those who remain in the countryside, rather than migrate to overcrowded cities where there are no jobs, the prospects may also be bleak. Agricultural economist Riad El-Ghonemy, senior associate at the International Development Centre, University of Oxford, and author of *The Dynamics of Rural Poverty*,[28] is convinced that the market-oriented neo-classical economic approach reflected in the GATT agreement "loses sight of people, the totality of the institutions governing their functions, and the very real world they live in."[29] He is particularly critical of its effects on the rural poor in the developing countries of the Third World:

The overriding concern of the neo-classical approach is with output growth and market determined prices. The problem with this approach, however, is that it disregards the distributional consequences—and the cost to food security and the net buyers of food among the poor. . . . Where the distribution of land ownership and opportunities is highly skewed, the market works for the benefit of those already wealthy, the traders, large- and medium-sized farms, and multinational corporations.

Take Malawi, where privatization of the food-producing communal land coupled with the free play of market forces led to land ownership becoming concentrated in the hands of merchants and traders who, looking for the most profitable return, planted tobacco for export rather than traditional food for local consumption. This led of course to great food insecurity on the part of the poorest sectors of the population. What this policy of export-led growth wrongly assumes for the least developed countries is that the growth in national income will actually reach the poor at an adequate rate.[30]

In short, where the rural poor are concerned, El-Ghonemy has little faith in the Reagan era's famed "trickle-down" economic scenario. A more likely denouement, some believe, would be the reestablishment of international agricultural trade patterns common to the pre-Second World War colonial system, a kind of repossession of former colonies, not by military force, but through economic pressure. Wrote Lang and Hines:

The GATT vision is for agriculture to be more export focused. Historically, as it did in colonial times, this has often meant that industrialized economies make the bulk of the money from finally processed products, such as coffee and chocolate.[31]

Declining Industry

It can be argued, of course, that this socio-economic picture is not only alarmist, but in the long run irrelevant, since statistics show that primary agriculture is a sector whose importance is inevitably declining relative both to production of processed foods and to the globe's total economy, and whose less-efficient practitioners would be better off doing something else.

There is no doubt that agriculture's share of world economic activity is shrinking. As the FAO reports:

> Trade in agricultural products has tended to lag behind trade in other sectors, particularly manufactures, as industrialization proceeds. On a global basis, agricultural exports now account for less than 10 percent of merchandise exports, compared to about 25 percent in the early 1960s.
>
> The tendency for agricultural trade to lose relative importance in external trade has been common to all regions, but in the developing country regions the process was particularly pronounced.[32]

A graph published with this report, depicting "agricultural as percentage of all exports," shows a sharp drop from roughly $300 billion in 1961 to less than $80 billion in 1993.[33] As for the shares of bulk versus processed food products, an Agriculture and Agrifood Canada report shows the world agri-food trade is definitely "changing significantly from bulk to more highly processed consumer oriented products." Trade in bulk products dropped from roughly 47 percent in 1983 to 36 percent in 1992, while trade in processed consumer products rose from approximately 31 percent to nearly 43 percent in the same period. On a graph, the trend lines form an X.[34]

It is tempting for those seeking simple answers to conclude that primary agriculture's decline in value is in fact inevitable—even desirable—and that planners ought to focus their efforts on areas of "growth." Such a rush to judgment, however, would ignore, as Riad El-Ghonemy put it, "the distributional consequences," expressed in terms of populations pauperized, starved, and forced into mass migration. For it is precisely the declining value of the raw agricultural materials they produce, from coffee to jute, that goes so far to keep Third World farmers poor. Also contributing to this poverty is the increasing share of the world market claimed by processed food and other value-added products made in the northern industrial countries by already established industries whose comparative advantage can only increase under the GATT regime. The process is rendered inevitable not by some blind working of fate, but by economic forces put in place as a result of trade negotiations.

Accepting the statistical trend as inevitable also ignores the dangerously shallow, exclusionary nature of the economic "givens" on which theorists—Ghonemy's "blinkered economists trapped in their own ideologies"—base their advice to governments. An *Atlantic Monthly* article, for example, put such economic yardsticks of growth as Gross Domestic Product (GDP) in proper context:

> The GDP is simply a gross measure of market activity, of money changing hands. It makes no distinction whatsoever between the desirable and the undesirable . . . it looks only at the portion of reality that economists choose to acknowledge . . . the GDP not only masks the breakdown of the social structure and the natural habitat upon which the economy—and life itself—ultimately depend; worse, it actually portrays such breakdown as economic gain. . . .
>
> When one considers the $32 billion diet industry, the GDP becomes truly bizarre. It counts the food that people wish they didn't eat and then the billions they spend to lose the added pounds . . . the coronary bypass patient becomes almost a metaphor for the nation's measure of progress: shovel in fat, pay the consequences, add the two together and the economy grows some more.
>
> So too, the O.J. Simpson trial. When the *Wall Street Journal* added up the Simpson legal team . . . network news expenses, O.J. statuettes and the rest, it got a total of about $200 million in new GDP, for which politicians will be taking credit in 1996. "GDP of O.J. trial outruns the total of, say, Grenada," the *Journal*'s headline writer proclaimed.[35]

An intellectual system that places the "growth" created by the circus show-trial of a celebrity accused of murdering his wife higher in its value-scale than the total production of real goods—including primary food products—of entire countries, is far from being rooted in reality.

Nor does it take into consideration many side-effects that are of crucial importance to people—human health, for example—which are considered merely peripheral by mainstream economists. An example was the scandal caused in the spring of 2000 by the contaminated-water deaths of residents of Walkerton, Ontario—deaths that were subsequently tied to the economically right-wing provincial government's decision to gut its environmental inspection facilities. Even the business-oriented Toronto *Globe and Mail* was shocked:

> Make no mistake: the Mike Harris government's environmental policies contributed to the deaths in Walkerton. . . . Two paths converged in Walkerton: government downsizing and municipal reform. When they came into office in 1995, the Harris Conservatives vowed that their Common Sense Revolution would make government smaller and more efficient. In the Environment Ministry's case, this meant, among other things, slashing the ministry's budget by 40 percent over five years, while taking away many of its (water inspection) responsibilities.[36]

For more on the environmental impact of neo-conservative economics, See chapter 4.

Finally, while the percentage of people engaged in farming may be declining, 55 percent of the world's people still live in rural areas and some 43 percent are actively engaged in farming.[37] To dismiss more than half the earth's population as economically marginal is madness.

If journalists are to perform their basic function, not only to "entertain" but to "inform and educate," it is incumbent upon them to report the economic movements affecting agriculture and the world's rural people in realistic rather than ideological terms, terms that take into account—quite literally—the situation on the ground.

CHAPTER FOUR

Killing the Goose:
Farming and Environment

*Just consider that U.S. farmers working 500 to
700 acres have smaller net incomes than
Japanese farmers on three to five acres.*[1]
—Masanobu Fukuoka

His name was Leduc and he was, unfortunately, our neighbor on the
"Rang du Quarante," in the heart of Quebec's dairy country. Our
farm was a small, part-time operation devoted to cash cropping hay
and oats for local dairyfolk, most of them fourth- and even fifth-
generation farmers whose awareness of the land, weather, and
wildlife was so ingrained as to be almost a sixth sense. They loved
the land and knew their place on it. Most milked herds of 20 or 30
Holstein-Friesians, sometimes brown-spotted Ayrshires ("Anglo
cows"), on mixed farms of 100 or 200 acres, including orchards,
sugar bushes, chickens and ducks, and the odd riding horse. They
didn't get rich, but took care of their families, put the kids through
school, and saved enough for retirement.

Leduc was different: like Caesar, he was ambitious.

When a neighbor whose children didn't want to farm
retired, Leduc bought the milk quota[2] and land, doubling his own
holdings overnight. He installed tile drains in the new fields, built
an open-concept feed barn where his cattle could be fed without
resort to pasture grazing, and bought a new tractor—a big White

"prairie-pounder," designed for use on Western Canadian wheat farms, that towered over him and could haul massive new machinery behind it. To make room for the tractor and machinery to turn around unimpeded, and maximize the area planted to crop, he razed the trees along the fencelines between his new and old fields, and filled in the ditches that divided them. Then he bought more land.

To pay for the new quota, barn, cattle, land, and machinery, he took out loans, on which the interest alone equalled the yearly profit of some of his smaller neighbors' operations. To meet the payments, he put all his fields to work growing high-yielding hybrid maize, a fast cash-return crop whose profits, with his milk sales, could be quickly funneled into his creditors' pockets. He was running what government extension agents advised was an efficient farm, and hoped to make it still bigger and more efficient.

Our place bordered his land on two sides, and we soon found what the real meaning of "efficiency" was. The first sign was the half-empty fertilizer sacks, tossed into the stream between our properties. Maize, especially modern, high-yielding hybrid varieties, is a hungry crop, drawing large amounts of nitrogen from the soil. Traditional farmers rotated the crops in their fields frequently, alternating maize with nitrogen-fixing plants like clover or beans so as to replace what the maize had taken. Rotating also prevented maize pest outbreaks, by eliminating the host plant for certain periods and disrupting insects' life cycles. But Leduc couldn't afford that. Maize brought the highest fast-cash return, so he kept on planting it, year after year, in the same fields. To replace the lost nitrogen and other soil nutrients, he purchased expensive chemical fertilizers and spread them on his fields in ever-increasing amounts. To deal with weeds and pests, he bought equally expensive chemical herbicides and pesticides. Time was money, so the distribution of fertilizer, herbicides, and pesticides was done in a hurry, with large machines and temporary hired help, and the sacks discarded carelessly.

By now, Leduc's herd had tripled and was housed and fed almost entirely indoors. Unwanted manure from his huge barn, more than he could possibly use on his land, piled up outside. Rain liquified it, carrying it away in runoff water.

There were fish in the stream when we bought our place, pumpkin seeds, sunfish, and a few suckers. The odd blue heron would hang around, looking for lunch. But after Leduc's hired men began dumping used sacks there, and the manure pile began leaching into it, the fish died, and the herons stayed away. The water took on a turbid, greasy look.

What happened to the land was worse. The soil in the river valley we farmed was heavy blue clay, difficult to drain, and took special care to maintain its tilth and friability—its structure. The layer of good, nutrient-rich topsoil was thin. Under traditional farming methods, growth of soil organisms—bacteria, earthworms, and helpful insects—and retention of soil nutrients was encouraged. Cultivation was rarely overdone. But Leduc's heavy machinery packed down the soil, creating a hardpan layer just below the depth of a plowblade. Overuse of chemicals killed off soil organisms, "burning" the powdery topsoil layer, which got thinner every year as the wind, now unbroken by fencerows, blew it away.

I walked onto my ambitious neighbor's property one day to look it over, and bent to touch the earth. It was no longer black, but grey, with the consistency of dry, fine gravel. What Leduc was doing, I told a friend later, was not farming. It was outdoor hydroponics, the soilless growing method used in some greenhouses. After a few years of this treatment, the soil was dead, a mere physical solid for plant roots to grip to keep their stems upright, while all the nutrients were purely chemical, washed through the field in solution. It was the most unnatural thing I'd ever seen, and couldn't possibly have continued many more years before the excessive costs would outrun Leduc's ability to continually borrow and expand. And when it all collapsed, there would be nothing left to sell but a flat stretch of lifeless grey gravel, where once had been good farmland.

In the late 1970s we sold our own place and moved away, and so never had to witness the final scene.

Our ex-neighbor's story was an extreme case, at least in the speed with which it occurred. But it was far from exceptional in basic philosophy or the quality of its environmental effects, as other investigators have shown.

Gordon Conway and Jules Pretty summarize the situation:

> Industrial activity has always resulted in pollution. But agriculture, for most of its history, has been environmentally benign. Even when industrial technology began to have an impact in the 18th and 19th centuries, agriculture continued to rely on natural ecological processes.

Crop residues were incorporated into the soil or fed to livestock, and the manure returned to the land in amounts that could be absorbed and utilized. The traditional mixed farm was a closed, stable and sustainable ecological system, generating few external impacts.

Since the Second World War this system has disintegrated. Farms in the industrialized countries have become larger and fewer in number, highly mechanized and reliant on synthetic fertilizers and pesticides. They are now more specialized, so that crop and livestock enterprises are separated geographically. Crop residues and livestock excreta, which were once recycled, have become wastes whose disposal presents a continuing problem for the farmer. Straw is burnt since this is the cheapest and quickest method of disposal. Livestock are mostly reared indoors on silage on farms whose arable land is insufficient to take up the waste.[3]

This postwar, capital-intensive way of farming matured first in North America and Western Europe, where petroleum products needed to fuel heavy machinery were relatively cheap and where the chemical firms that manufactured the required fertilizer, pesticide, and herbicide were centered. It was rapidly accepted as the norm by most university agriculture faculties and government extension services because it was an "instant hit," producing bumper crops and huge surpluses of corn, wheat, and other commodities. It was assumed as the basic model by the scientific plant breeders who, in the late 1950s and early 1960s, launched the so-called Green Revolution, which carried the industrial farming model around the world—in some cases imposing it as an aid condition on Third World governments.

(For those unfamiliar with its story, the Green Revolution began as a publicly funded research effort aimed at applying science to the food problems of developing countries. The International Rice Research Institute (IRRI) in the Philippines and the International Centre for Maize and Wheat Improvement (CIMMYT) in Mexico united the efforts of research scientists, foremost among them Norman Borlaug of the United States, in developing new plant varieties that would produce higher yields to meet the food needs of rapidly rising populations.[4] In many ways, the fruits of the research at these and other centres proved beneficial. Enormously productive crop varieties were developed and exported to countries such as India, where—grown in monocultural systems—they proved especially successful in irrigated lowland environments. Once a net grain importer, India became self-sufficient in grain thanks to the new varieties, which by 1990 accounted for "almost 70 percent of the combined rice, wheat and maize area in the developing world."[5] So positive were initial results that aid agencies like the World Bank became enthusiastic proponents of both the crops and farming systems associated with them.)

A Flip Side

Like other hits, however, industrial farming has a "flip side" that, as shown in the case of farmer Leduc, includes six categories of undesirable environmental impacts: pollution, wildlife habitat destruction, soil degradation, waste of freshwater resources, loss of biodiversity, and threats posed by introduced or "exotic" species (including those produced via the gene manipulation of so-called "bioengineering").

Pollution

Conway and Pretty provide clear descriptions of the pollution problems created by industrial farming:

> The primary environmental contaminants produced by agriculture are agrochemicals, in particular pesticides and fertilizers. These are deliberately introduced into the environment by farmers to protect crops and livestock and improve yields. Contamination is also caused, though, by the various wastes produced by agricultural processes, in much the same way as occurs in industry. The wastes comprise straw, silage effluent and livestock slurry, and, in the Third World, the wastes from on-farm processing of agricultural products such as oil palm and sugar. From the immediate environment of the farm

contamination spreads to food and drinking water, to the soil, to surface and groundwaters and to the atmosphere, in some instances reaching as far as the stratosphere.[6]

The worst offenders are the wide variety of herbicide and pesticide compounds, whose residues can enter the food chain, causing cancers and other diseases in livestock and humans, and impacting the natural environment in myriad ways. For example, two of the best known herbicides, the compounds 2,4,5-T and 2,4-D, both contain the highly toxic contaminant dioxin (2,3,7,8-TCDD), a chlorinated hydrocarbon created as a byproduct during manufacture.[7] One of the most deadly substances known, dioxin was present in the infamous Agent Orange herbicide used by American forces in Vietnam and believed responsible by many researchers for thousands of abortions and birth defects among humans exposed to it.[8]

Impacts on beneficial insects can be equally severe, as in the case of honeybees. According to Conway and Pretty:

> The first modern insecticides, the organochlorines, were relatively nontoxic to honeybees, but many newer organophosphates and carbamates are very hazardous. During the 1970s pesticides annually destroyed 40–70,000 bee colonies in California, some 10–15 percent of the total, while the annual national loss of colonies was estimated at half a million. At lower doses, pesticides may increase the aggressiveness of bees. . . .
>
> Alfalfa is a crop particularly attractive to both wild and domestic honeybees. In one incident in Washington, large numbers of bees were killed by diazinon and two years later had only regained a quarter of their previous population level.[9]

When it is realized that honeybees are not only the source of a honey industry worth millions of dollars, but also necessary for pollination of farm crops from orchards to clover, the seriousness of such losses becomes evident.

Herbicide and pesticide contamination are particularly high where intensive, industrial farming is common: U.S. states like Iowa, Minnesota, and Ohio, where intensive corn and soybean cropping is practiced. It is also high in many Third World countries, where compounds like DDT, dieldrin, and aldrin, outlawed in northern countries, are often "dumped" and where both herbicides and pesticides may be applied without proper training or operator protection.[10] It is estimated that "about 40,000 people in the developing world die of pesticide poisoning every year."[11]

Industrial farming favors, even demands, use of toxic compounds, as the authors of *Unwelcome Harvest* explain:

> In some parts of the world, pest problems first became serious with the expansion of irrigation and the increased use of chemical fertilizers. Year round irrigation makes double cropping possible and if the same crop is grown continuously, explosive pest outbreaks can occur. In general, the new cereal varieties tend to be more susceptible to pests and diseases and the loss of crop heterogeneity has favored high pest and disease populations.
>
> New practices, such as the direct sowing of rice, have led to increased populations of grass weeds, requiring more herbicides, while greater use of nitrogen [fertilizer] has heightened susceptibility to diseases. Often new practices come as packages of interlinked components. Direct seeding, for example, requires support from intensive herbicide use. This way, farmers can become locked-in to an intensive system where pesticides appear to be indispensable.
>
> Mechanization can also trigger pest problems. In the vegetable growing region of the Thames Valley, in the UK, mechanization has produced economies of scale leading to large farms that concentrate on growing only three types of vegetable. There are fewer crop rotations and farms are essentially monocropped with higher-yielding varieties susceptible to pests and diseases. . . . As a result, pesticide use has increased dramatically.[12]

Government subsidies aimed at boosting production, especially for export, also tend to foster greater chemical use. In some Third World countries, pesticide subsidies are as much as 89 percent of the retail cost of compounds.[13]

In addition to toxic agents intended to kill unwanted insects or plants, inorganic chemical fertilizers—mixes of nitrogen, phosphorus, and potassium (NPK)—pose pollution dangers when used excessively. Nitrates and phosphates not taken up by crops enter the soil water, and from there the water supply of animals and humans downstream from farms. When the nitrogen level in drinking water (as nitrate plus nitrite) exceeds 10 mg/L it can interfere with oxygen transport within the body. According to Agriculture Canada:

> Children under one year old are particularly susceptible, as are cattle and young animals. There have been cases where excessive nitrate in the water of farm animals has reduced conception rates and decreased the number of live births.[14]

Another unwanted side-effect of heavy fertilizer use is eutrophication of streams, ponds, rivers, and other water bodies. Eutrophication—oversupply of nutrients—can cause rapid, heavy growth of algae and other plants, which ingest oxygen and give off carbon dioxide. As a result, the underwater environment becomes choked with plants and the oxygen balance is disturbed, leaving insufficient oxygen for fish and other creatures. Massive fish die-offs can occur after such contamination. Dense growths of algae associated with eutrophication can cause a further threat, from toxic compounds secreted by such algal species as *Prymnesium* and others. Fish and livestock—particularly pigs—are vulnerable to various algal toxins.[15] Phosphorus stimulates the growth of blue-green algae, and agriculture is estimated to contribute approximately 40 to 60 percent of the phosphorus entering the Great Lakes through tributary rivers.[16]

In addition to these problems, serious difficulties are posed by creation of animal waste in industrial farming. The size of the problem was first suggested 30 years ago by the American Association for the Advancement of Science (AAAS):

> As recently as 15 years ago, a one-man dairy operation included about 20 cows; today it has 50 to 60. Confined housing, well-engineered and executed layouts with good traffic patterns, automatic feeding and milking, and mechanical materials-handling equipment make it possible for a dairyman to enlarge his operation. There are farms with hundreds and even thousands of cows in production. The amount of wastes to be handled can

vary anywhere from 100 to 200 pounds per cow per day
. . . the dairyman from a 100-cow dairy operation will
have to dispose of at least five tons of wastes per day during all the 365 days of the year (366 days in a leapyear),
or something like 1,825 tons in a year's time. . . .

Nearly all beef animals spend one-third of their life-
time, three to four months, in feedlots, where they are
fattened and polished before they are marketed. Waste
disposal problems from the feedlots, even assuming they
are well managed, arise from the large number of animals
in each lot (anywhere from 500 to 5,000 head of cattle)
and thus the large quantities of manure involved with no
land to spread it on.[17]

Poultry farms cause even greater problems:

In the major poultry producing regions, about 50 to 80
percent of the egg-laying hens and almost all broilers are
raised under confinement. Practically all large poultry
operations are highly mechanized, with automatic feed-
ing and watering, optimum ventilation and temperature
control, and automatic egg collection; but few have adequate waste disposal systems. The number of birds on
some poultry farms exceeds one million. On the basis of
a population equivalent of 1.7 per 100 birds, the wastes
from one million chickens would be equivalent in
strength to the wastes from a city of 68,000 people. The
managing and proper disposing of the wastes from a city
of 68,000 people is a formidable task. . . . Several poultry operations in several states, some with investments of
more than $100,000, have been forced out of business
because of unsatisfactory systems for waste management
and disposal.[18]

Manure represents a significant health threat to humans living downstream from farms. For example, the infectious bacteria
salmonella can survive for up to a year in liquid manure and is easily transmitted to people. According to Agriculture Canada, other
infections transmitted to humans via manure include: "anthrax,
tularemia, brucellosis, erysipelas, tuberculosis, tetanus and colibacil-

losis."[19] In the years since the AAAS symposium, livestock operations have grown still more concentrated and specialized, and disposal problems have increased accordingly.

This trend came to a horrifying culmination in the spring of 2000, when an outbreak of E. coli bacteria in the drinking water of the Canadian town of Walkerton, Ontario, killed at least six people and made an additional 2,000 ill. A report submitted to the Ontario legislature by Environmental Commissioner Gord Miller pointed the finger at the region's industrial farms. *The Canadian Press* wrote:

> Coinciding with a public inquiry into the water-borne E. coli outbreak that infected 2,000 people and contributed to six deaths at Walkerton, Miller's report criticized lack of regulation of large farms that produce vast amounts of liquid manure.
>
> He said the situation has made Ontario a "haven" in North America for companies wanting to avoid rigorous controls.
>
> "In many other jurisdictions, including Quebec and the U.S., there are laws and regulations governing the management of manure. But in Ontario there is virtually no control," he said.
>
> The report proposed tightening industry standards and subjecting large farming operations to an approval process.[20]

A graphic description of the "large farming operations" in question was provided in another article on the crisis:

> Born and raised on one farm, raising her kids today on another farm, Anita Frayne was no agricultural activist—until a neighbor built a barn that holds 3,400 hogs.
>
> And until that neighbor—actually a corporate owner—let the pig manure flow through pipes underground into a creek, and into nearby Lake Huron.
>
> "I've stood on the beach and smelled pig manure," she said. Sometimes liquid manure runs out through a creek. "It's brown. It stinks. You smell pig manure and your nose will lead you to it."[21]

Wildlife Habitat Destruction

The threat posed to wildlife by modern farming is not limited to insects and the underwater world, nor traceable only to pollution. Of greater concern is the loss of habitat as more land is brought under cultivation, and land already cultivated is consolidated in ever larger monocrop operations.

In contrast to natural forest, which seen from the air presents a relatively solid expanse of green, regions given over to farming are characterized by "mosaic landscapes," an irregular checkerboard pattern of varicolored fields and woodlots connected by zigzagging fencelines. For insects, birds, and mammals on the ground, the helter-skelter squares constitute a complex jumble of micro-environments, with fencerows linking them.[22] Under traditional mixed farming regimes, the jumble was complicated enough that it often benefited wildlife, providing shelter and/or forage for different species at different seasons. In North America, for example, the expansion of agriculture in northern New England and parts of Canada caused a decline in woodland moose populations, but resulted in a veritable population explosion among whitetail deer—which eat maize and find the open, mixed-farm environment ideal. The same has been true of various insect, rodent, and bird species.

The landscapes created since the Second World War by industrial farming represent a marked, in some cases radical, environmental simplification, with negative results for most wildlife species. Britain provides an example of the effects on birds, examined in *Farming and Birds*:

> Mixed farming systems, combining tillage crops and stock, frequently emerge in our analyses as most favorable to birds. Yet farming has changed dramatically over the last 40–45 years, from an industry whose methods were relatively favorable to wildlife to a specialized, highly technical business heavily biased against the maintenance of the diversity of nature . . . "Modern farming is the antithesis of the image projected by the old adage about sowing corn—one for the rook; one for the crow; one to rot and one to grow. Nowadays the farmer sows [only] one to grow; competition within the industry allows little margin for natural wastage." (Wright, 1980).[23]

51

Farming affects bird populations in two ways:

> The first is by engulfing and totally destroying particular habitats within farmland, such as woodlands and lowland heaths. The second is by modifying the nature of the surviving habitats, thereby altering the niches they offer in ways that make them less (or perhaps occasionally more) attractive to birds. . . . Since 1939 the conversion of permanent pasture to arable production has constituted a major loss of habitat.[24]

Engulfing usually occurs when the land is first cultivated. In Britain, this may have been hundreds of years ago. Industrial farming's much more recent effects are felt in the form of "modifying the nature of surviving habitats," though in some cases "modifying" should be replaced by "eliminating." The changes are most often due directly to mechanization and farm expansion. O'Connor and Shrubb write:

> For modern arable farms to be viable with low labor inputs, farm structure must allow the most efficient use of large machines. The impact of such machinery on farmland practice [is] magnified by the growth in the average size of farm holdings in Britain. Between 1875 and 1979 farm sizes have increased by a factor of 2.0–2.5.[25]

As for the effects on birds:

> the increasing size and speed of machines, which tend much more to divorce the operator from his immediate surroundings, may well increase nest and brood losses in ground-nesting species. Second, and probably more important, are the effects of mechanization in narrowing the complexity and timing of crop rotations.[26]

On a traditional farm, where as many as six to ten field crops may have been combined with garden vegetables, pasture, and livestock, crop rotations were frequent, often complex. As O'Connor and Shrubb note, "the classic four-course Norfolk rotation [red clover (*Trifolium pratense*), winter wheat, sugarbeet, and spring barley] not only provides cleaning crops (roots) and compost or manuring crops (clover) but also provided for the most economic use of labor by spreading the farm's work fairly evenly throughout the year."[27] Under industrial farming, however:

> the bulk of the crop is now planted in autumn, despite the workload this imposes. Formerly the work used to be divided fairly evenly between spring and autumn. The resulting change in the timing of cultivation has also been promoted by the decline in root crops in many areas and the increasing concentration on cereals. . . .
>
> changes in the periods of tilling the soil have a significant impact on the availability of food supplies to birds, particularly in the breeding season. Spring cultivations are a major source of invertebrate food for many birds when the start of breeding may be limited by the availability of food. Autumn-sown fields are unsuitable feeding areas in spring since the vegetation is by then too tall and dense. Ground-nesting birds such as Lapwing may also prefer spring cultivations for nesting.[28]

The removal of fencerows to consolidate farm fields also destroys bird habitat, and cuts off the sheltered connections between fields along which rodents and other small mammals—which make up the diet of most hawks and owls—travel. Drainage of wetlands, and filling in of drainage ditches between fields, reduces habitat for ducks and other waterbirds. An example of how great the impact of human activity on birds can be is the greater prairie chicken. Once common on the western plains, it has been extirpated from Canada since the turn of the century, largely due to habitat loss caused by human settlement.[29]

What is true for birds is true for other species, in locations as far apart as Western Canada and East Africa.

On the Canadian prairie, writes David Wylynko:

> Loss of habitat [due to agriculture] is cited as the primary reason for the decline in most prairie species. During the 1900s, four mammals—the grizzly bear, wolf, black-footed ferret, and swift fox—have been extirpated from the Canadian prairies. Three more—the bob-

cat, long-tailed weasel, and badger—are at considerable risk.

Both the number and abundance of bird species have decreased dramatically in the transition from native habitat to cultivated land and tame pastures as well. Among birds that breed in or migrate through the prairies, more were placed on Canada's list of species at risk in 1994 than in any other ecozone in the country.[30]

Modern agriculture has been identified as an indirect threat to the future of the burrowing owl and prairie falcon, because of farmers' dislike for these species' main food source: Richardson's ground squirrel. Ground squirrels, or "gophers," are considered vermin and exterminated by farmers because their varied diets include wheat, barley, and a number of popular garden vegetables, while their burrows pose dangers for cattle that step in them. Though not themselves endangered, the gophers' elimination destroys a forage source for much rarer predators that eat them.[31]

In Africa, farming is likely the most serious potential danger—more serious in the long term than that of today's ivory poachers—to *Loxodanta africana*, the African elephant. While international attention is focused on high-profile anti-poaching "wars" and public burning of impounded ivory, the threat to elephant survival posed by habitat loss due to the expansion of agriculture and grazing goes comparatively unreported. Yet the threat is clear, as one recent statement of the economic aspects of the problem reveals:

> The competing for land argument embraces Hardin's notion of the survival of the fittest, where the demands of two sympathetic species are sufficiently similar that competition between them leads to the extinction of one. In this case, the species are humans and elephants, who compete for essential resources of food and habitat. Humans' plant food demands are similar to those of the elephant, and they also indirectly compete for the use of the same resources for their domestic stock. This competition is likely to be significant, given that the African population doubling-time is now merely 18 years, and has brought about rapid forest conversion for agricultural and pastoral activity.[32]

This basic economic cause for competition between the two species is accentuated by the tendency of Africans not to consider any long-term benefits that might accrue from the conservation of wildlife. As the study authors note:

> Rural people in Africa . . . are likely to prefer income in the present as opposed to the future. In Africa, average life expectancy at birth is 51 years and infant mortality is 10 percent. These figures are even more grim in rural areas where health care, nutrition, education and clean water are scarce. Thus risk of death is a crucial factor. Uncertainty of the future is also compounded by threats of drought and other natural disasters, political instability and warfare, economic disruptions and policy changes and natural resource degradation. Finally . . . under conditions of extreme poverty, a household's major concern is securing sufficient means for survival today. Observers suggest these conditions leading to a high rate of time preference are reflected in the lack of sustainable farming practices (practices which conserve soil quality, irrigation water and rainfall) in sub-Saharan Africa. This implies that individuals are not willing to bear the risk and uncertainty from changing farming patterns in order to benefit from future gains.[33]

Nor are they willing to accept the present checks of being kept out of game areas when they want to expand their land holdings, or of elephants destroying crops on already-cultivated fields. As a villager in Zimbabwe once observed: "Wildlife is nothing but a nuisance. Elephants destroy our crops every night. They (the government) can kill everything bigger than a hare as far as we are concerned."[34]

A solution to the conflict exists—one that protects both elephant herds and the financial interests of African farmers and rural villagers—and has been successfully field-tested in Zimbabwe. But whether the Communal Areas Management Program for Indigenous Resources (Campfire) will be widely adopted by African governments is in doubt.[35]

Launched in 1986, the program is based on two principles: (1) the concept that wildlife is an economic resource with real value to the rural economy, and (2) the idea that local resources should be controlled locally. In the past, most African governments followed the game conservation pattern of colonial times, that of setting aside reserves for wildlife hunting and tourism, and keeping local rural people away, sometimes even evicting them from parklands. As African populations grew, and pressure increased to put more land—even marginal land better suited for wildlife—under cultivation, rural people began to resent being barred from reserves. Squatting—illegal occupation of land—and poaching increased in some countries to the point of "virtual warfare" between rangers and local villagers.[36]

Realizing a crisis had been reached, Zimbabwe tried a new approach. Management of wildlife in communal land areas was given over to local district councils, which adopted a quota system. An optimum number of wildlife that could be supported in a given habitat was set, and surplus animals, as well as those that caused damage to nearby crops, were culled. The meat, hides, and ivory were sold and earnings were used partly to pay for wildlife management (including anti-poaching patrols) and partly returned to local villages for use in such development projects as school and hospital construction, and for cash distribution to local households. As the deputy director of the Zimbabwe Department of National Parks and Wildlife Management notes:

> Government has issued a set of guidelines to councils suggesting that a minimum 50 percent of total income should be returned to the wards and villages where it was generated. A maximum 35 percent should be reinvested in wildlife management costs for the following year and a maximum 15 percent should be retained by the council for administrative purposes . . .
>
> From recent studies of the wildlife industry in Zimbabwe, it seems the net financial returns from land under wildlife significantly exceeded those possible from cattle (US$1.11 per hectare versus $0.60 on commercial farms in Natural Region IV of Zimbabwe), and the potential for improvement in wildlife returns is far greater than that for cattle (up to $5 per ha for hunting and up to $25 per ha for ecotourism). For the past 20 years the amount of land allocated by landholders to

wildlife has been increasing and, including state-protected areas, almost one-third of Zimbabwe is now under wildlife. The trend is likely to continue as marketing of wildlife improves.[37]

As a result of this innovative program, the elephant herd in Zimbabwe actually increased during the same years that the herd in nearby Kenya, with a protectionist philosophy, plummeted dramatically. Despite its success and popularity with farmers, however, the Campfire program has been attacked in Zimbabwe and abroad. Internally, R. B. Martin writes:

> Campfire is seen as a threat: rural people have organized themselves into suitable groups and established committees to undertake wildlife management. These institutions are becoming vociferous in the political arena. If their representatives, both on local committees and at the level of members of Parliament, don't "produce the goods," others are elected in their place. Unexpectedly, Campfire has become the engine for a powerful democratic movement among remote communal land people. As such it is a threat to politicians and bureaucrats who do not really wish to see self-sufficient rural communities.[38]

Outside Africa, the program—because it permits hunting and the sale of ivory—is viewed by some environmental and animal rights groups as heresy:

> Certain extreme right wing international "green" organizations perceive the program as a threat to their conservation ideologies. If it can be demonstrated that wildlife populations increase through sustainable use programs, this weakens international stands against any form of wildlife exploitation by humans.[39]

An international effort has been launched in the northern industrial region to discredit the Campfire approach, and many donor countries are reluctant to offend constituents by allocating funds for similar efforts in other African nations. The eventual issue of this highly newsworthy battle of conservation philosophies—a battle that is essentially a "farm story"—could well decide the fate of the African elephant.

Soil Degradation

Industrial farming, especially when carried out carelessly and without due regard to local environmental characteristics, can do great harm to soils, promoting topsoil erosion, soil compaction, salinization, and loss of fertility. Agriculture Canada researchers warn the damage is often not immediately apparent:

> Problems develop only over long periods. The short-term, negative effects of various farm management practices on the soil are often hard to identify. In the long-term, cumulative effects reduce soil productivity. When nutrient-rich topsoil is lost, the problem can be overcome in the short term by increasing fertilizer input. However, this adds to production costs. Other forms of degradation are less easily remedied. For example, using more fertilizer does not increase yields if plant growth is limited by compacted soil layers hindering root development.
>
> Besides the negative effect experienced by the farmer, soil pollution and soil degradation also have long-term implications for society. Not only is the land's potential for food production reduced, but other large-scale environmental impacts, such as the spread of desert lands and flooding, could occur.[40]

A number of practices common in industrial farming—excessive tillage, summer fallowing, and razing windbreaks to permit expansion and consolidation of fields—increase wind and water erosion, causing various noxious effects:

> Besides contributing to pollution, soil eroding onto other properties creates a nuisance. It can also clog drainage and irrigation channels. Wind and water erosion removes the fine topsoil and associated nutrients from farmlands. If not controlled, it decreases the pro-

ductivity of land as the subsoil content of the plowed
layer increases. As subsoil is incorporated into the cultivated layer, soil fertility and water-holding capacity are
reduced. Root growth and development are also limited,
resulting in variable crop growth. Soil transported by
erosion and redistributed within the field can increase
localized ponding, smother young seedlings, and cause a
crust to form on the surface after drying. In severe cases
erosion produces such extensive gullies that workable
land is lost completely.[41]

Eroded soil that enters waterways becomes sediment that
eventually fills in shipping channels, clogs hydroelectric dams, and
causes fish die-offs. In the Great Lakes basin, dredging of harbors to
clear them of silt and keep shipping lanes open costs in excess of
US$100 million annually.[42]

Soil compaction, caused by poor tillage practices—including use of big machines whose weight compresses the soil (a pen-
chant of farmer Leduc)—brings its own problems:

A few soils are naturally dense and limit plant growth.
Man can create similar, undesirable conditions by working soil when it is wet and using machinery with excessive weight or at excessive speeds. Repeated cultivation
can also increase oxidation rates and microbial decom-
position. These phenomena reduce the organic matter
content, making it easier for tillage to pulverize soil
aggregates and destroy soil structure. Loss of structure
makes the soil more susceptible to erosion and further
compaction.[43]

Under an industrial agricultural regime, time is money, and
speed in performing tillage tasks is of the essence. Large machines,
operated at maximum speed by hired hands, race through their
work, with little attention to the long-term effect on soils. Debts
must be paid down every month, regardless of the biological needs
of soil organisms. The eventual effect on the source of life itself is
not reflected in the mortgage interest tables.

Irrigation, an indispensable tool of modern agriculture, can
also threaten soils through salinization. When too much water is
used to irrigate fields, the water table can be raised, causing soluble

salts in the earth to rise with it. When the water evaporates, the salts are left in surface soil layers. Warns Agriculture Canada: "Contamination of the soil with salts reduces germination and growth of many crops. Increased sodium levels cause deterioration of soil structure by inducing the formation of dense soil layers that restrict water movement."[44] Leached salts in groundwater can bring problems far from the farmland where they originated, increasing the salinity of rivers that may be the only source of drinking water for downstream municipalities.

Waste of Freshwater Resources

The globe has enough fresh water to supply the needs of everyone, but much of that water isn't available when and where needed. Sandra Postel, director of the Global Water Policy Project, notes:

> At least 20 percent of the renewable water supply—generated each year by the solar-powered hydrological cycle—is too remote from population centers to be of use. A large portion runs off in floods, and cannot be supplied reliably to farms, industries and households. After accounting for this unequal distribution of water in time and space, the picture of plenty is revealed to be an illusion. [45]

Clashes over water rights are becoming increasingly common, both within countries and internationally, and when fingers are pointed at those deemed responsible for shortages, farmers are inevitably—and justifiably—singled out: agriculture accounts for two-thirds of all water drawn from rivers, lakes, and aquifers. If the massive amounts of fresh water used in farming are not used wisely or actually degrade the water supply (e.g., through salinization), political and even military conflict can result.

Much of the water used for agriculture does, in fact, result in waste of the water itself and degradation of other resources. Every year, up to 10 percent of irrigated lands lose their productivity due to salinization caused by poor water management.[46] Damage on a much greater scale has been caused by the construction of large dams and diversion schemes on the world's major rivers, inspired in large part by agricultural demands. As Postel points out:

> Globally, water demand has more than tripled since mid-century, and it has been met by building ever more and larger water supply projects, especially dams and river diversions. Around the world, the number of large dams (more than 15 meters high) climbed from just over 5,000 in 1950 to roughly 38,000 today. More than 85 percent of large dams have been built during the last 35 years. Many rivers now resemble elaborate pumping works, with the timing and amount of flow completely controlled by planners and engineers. . . .
>
> Unfortunately, this massive manipulation of the hydrological system is wreaking havoc on the aquatic environment and its biological diversity. River deltas are deteriorating, species are being pushed toward extinction, inland lakes are shrinking, and wetlands are disappearing. . . .
>
> In the Aral Sea basin, site of one of the world's worst environmental tragedies, what was once the planet's fourth largest lake has lost half of its area and three-fourths of its volume because of excessive river diversions to grow cotton in the desert. Some 20 of 24 fish species have disappeared, and the fish catch—which totalled 44,000 tons and supported some 60,000 jobs in the 1950s—has dropped to zero.[47]

An even greater disaster has been visited upon Egypt, where dam construction was ironically intended to boost farm production, as well as provide electrical power:

> The Aswan High Dam on the upper Nile, completed in 1963 and fully effective by the 1970s, was intended to regulate the river's flow and provide hydroelectric power. The traditional fertility of the Nile Valley, however, was dependent on more than 100 million tons of silt deposited annually as the river flooded. The silt is now silting up the artificial Lake Nasser, forcing farmers downstream to rely on fertilizers and robbing local brick-

makers of raw materials; about 35 percent of Egypt's cultivated land is suffering from salinization. Deprived of nutrients, the fish stocks of the eastern Mediterranean are declining, while the Nile delta is being eroded steadily. Schistosomiasis, a disease common in newly irrigated areas, has spread explosively, causing debilitation and death. In 1990 between five and six million people were affected.[48]

At international level, decisions by governments as to how to use the water from rivers that feed not only their own but also neighboring countries have resulted in serious inequities. An example is the Ganges River, which flows through India and Bangladesh. As Postel notes:

> Bangladesh suffers from India's unilateral diversion of the Ganges at the Farakka barrage, and the failure, since 1988, to reach agreement on how to share the Ganges during the dry season. In 1993, the dry season flow into Bangladesh was the lowest ever recorded. As riverbeds dried up and crops withered, the northwestern region suffered greatly. The Ganges Kobadak Project, one of this poor nation's larger agricultural schemes, reportedly suffered US$25 million in losses.[49]

Other river systems where international conflict over freshwater use rights has reached crisis level include the Danube River basin, shared by 17 squabbling countries;[50] the Nile River basin, whose waters are disputed by Egypt and Sudan; and the Tigris-Euphrates River basin, over which Turkey, Syria, and Iraq threaten to come to blows.[51] In each region, agricultural use is a leading bone of contention.

Loss of Biodiversity

Under traditional agricultural systems, farmers keep some grain from every harvest as seed for the next planting season. In each ecological region, sometimes after centuries of trial and error, they have selected plant varieties suited to their all-round needs: those best adapted to local weather and soil conditions, most resistant to local pests and plant diseases, and most productive of the wide variety of products needed on a typical mixed family farm. They choose varieties of rice, wheat, barley, millet, or sorghum that produce not only enough grain for human consumption, but also straw and leaves for livestock feed, as well as for use as mulch or "green manure" on their fields. The number of available varieties is surprising. For example, FAO researchers in Ethiopia stopped to examine a local farmer's two-acre plot—and identified 11 varieties of indigenous wheat, which the farmer had planted for a range of specific uses.[52]

Of the world's estimated 500,000 plant species (roughly half of which have been scientifically classified),[53] farmers have domesticated hundreds that, when multiplied by the number of varieties developed within each species, provide a panoply of planting options capable of adapting to almost every environmental eventuality. As this author noted in a published interview with Murray Bookchin:

> Paraphrased, Bookchin's watchword might be expressed: in diversity there is strength. The dynamic tension maintained between a variety of organisms in an ecosystem gives the system its resilience. The more forms it takes, the less likely life is to be wiped out by fire, flood or Ice Age climate changes.
>
> The traditional mixed farm, producing a variety of crops, possessed the same strength. If hog prices were down, a good corn crop might take up the slack. An abundant maple sugar run could offset the loss of a Jersey calf.[54]

In modern, industrial farming, however, the goal of plant breeding is turned on its head: diversity as a trait is considered undesirable and deliberately "selected out." A single aim—most frequently to maximize grain production for immediate export—takes precedence over all others, and only crop varieties that contribute directly to it are favored. Plants with long stems, which might be useful on a mixed farm as straw, are discarded in favor of those with shorter stems and larger grain heads. Varieties that grow to the exact height required for fast, mechanical harvesting are favored over shorter or taller strains that might be resistant to local insects (it is assumed that pesticides will protect the more vulnerable plants). Farmers cease to save the seed of older, local plant varieties, which may as a result be lost forever. To assure processing factories of "uniformity of product," only the new varieties are planted—row on row, acre after acre—in huge monocrop operations:

> With everything invested in hundreds upon hundreds of acres of a single uniform crop, a tiny shift in market price would mean disaster if the system were not supported . . . by government intervention. Without the artificial protection of thousands of dollars worth of harsh chemical pesticides, the fragile monocultural crop could also be wiped out overnight by an invasion of insects attracted by the unnaturally rich feast spread before them.[55]

Such a major pest invasion—in this case a fungal infection rather than insects—has, in fact, already occurred, as Baeza-Lopez reports:

> The intrinsic weakness caused by the genetic uniformity of modern varieties was underlined in 1970, when the fungus *Helminthosporium maydis* attacked maize in the United States. Production dropped by 15 percent nationwide and 50 percent in the affected areas, causing losses of millions of dollars.
>
> The commission charged with investigating the maize disaster concluded the cause was genetic uniformity. Nearly all hybrid varieties in the country had been obtained from a single source of parental sterility called *Citoplasm texas*, which was susceptible to this new form of fungus.[56]

The lesson seems to have been lost on the industrial farm sector. Instead of a wide selection of local crops, the system continues to prefer the types of plants bred by the scientists who created the Green Revolution: high-yielding, "improved" grains. Most often these are hybrids that, because they are the offspring of genetically dissimilar parents, cannot "breed true" in subsequent seasons. Farmers cannot save the seed from such crops to plant again, but must purchase new stock each year from seed companies. Indian scientist Vandana Shiva analyzes the situation:

> The miracle of the new seeds has most often been communicated through the term "high-yielding varieties (HYVs)." The HYV category is central to the Green Revolution paradigm. However, unlike what the term suggests, there is no neutral or objective measure of "yield" on the basis of which the cropping systems based on miracle seeds can be established to be higher yielding than the cropping systems they replace. . . .
>
> The Green Revolution category of HYV is essentially a reductionist category which decontextualizes properties of both the native and the new varieties. Through the process of decontextualization, costs and impacts are externalized and systemic comparison with alternatives is precluded. . . .
>
> Since the Green Revolution strategy is aimed at increasing the output of a single component on a farm, at the cost of decreasing other components and increasing external inputs [e.g., inorganic fertilizers], such a partial comparison is by definition biased to make the new varieties "high-yielding," when at the systems level they may not be.[57]

Thus, a variety of wheat that produces abundant, large seed heads with a high protein content—but that has short, weak stems of little use as straw, is prone to lodging (wind damage), and is perhaps vulnerable also to local insect or bacterial or fungal pests—is labeled "improved." The fact that the variety may require large doses of costly inorganic nitrogen fertilizers and frequent applications of equally costly pesticides to survive—and that meeting these costs draws farmers into a debt spiral—is not considered relevant. Nor is sufficient attention paid to the fact that, in allowing local crop varieties to be discarded and die out, plant breeders are depriving themselves of the very sources of fresh germplasm their programs might need to meet as yet unforseen future needs. Shiva continues:

> The indigenous cropping systems are based only on internal organic inputs. Seeds come from the farm, soil

fertility comes from the farm and pest control is built into the crop mixtures. In the Green Revolution package, yields are intimately tied to purchased inputs of seeds, chemical fertilizers, pesticides, petroleum and to intensive and accurate irrigation. High yields are *not* intrinsic to the seeds, but are a function of the availability of required inputs, which in turn [may] have ecologically destructive impacts. . . .

In the breeding strategy for the Green Revolution, multiple uses of plant biomass seem to have been consciously sacrificed for a single use, with non-sustainable consumption of fertilizer and water. The increase in marketable grain has been achieved at the cost of decreased biomass for animals and soils and the decrease of ecosystem productivity due to overuse of resources.[58]

Critics of various Green Revolution projects stress the fact that the concentration on hybrid seed, which must be repurchased each year, leads to farmer dependency. One such project, conducted by Japan's Sasakawa Foundation in Ethiopia, has been described as:

an artificial laboratory. SG 2000 selects the crops and technical packages without farmer consultation, procures the inputs and delivers them to the participating farmers via the village extension agent who also handles credit recovery. The packages are high-input/high-output/high-risk and create farmer dependency on imported hybrid seed (Pioneer brand) and fertilizer.[59]

Farmer dependency becomes particularly ominous when it is realized that the international seed trade is becoming increasingly concentrated in the hands of a small number of companies—often the same chemical manufacturing firms that produce the fertilizer and pesticides that the new, "improved" seeds need to survive. Twenty years ago, most seed sold to farmers came from long-established, local seed houses. Today, more than 30 percent of the seed sold in the northern industrial countries is controlled by 20 large firms, many of them pharmaceutical or chemical producers. The top six in terms of sales are Pioneer Hi-Bred, Sandoz, Limagrain Inc., Nickerson, Upjohn, ICI, and Cargill.[60]

Still more sinister has been the corporate development, in collusion with the U.S. Department of Agriculture, of so-called "terminator" crops, whose seeds are genetically engineered *not* to reproduce if sown the following year. The technology has been widely condemned as "a dangerous and morally offensive application of agricultural biotechnology, because over 1.4 billion people depend on farm-saved seeds."[61] Many observers foresee famine if Third World farmers are forced by patent laws to depend on such crops, or if their genetic traits should accidentally transfer into other varieties.

The massive wastefulness that can result from corporate control of plant germplasm was further demonstrated when Seminis, the world's largest vegetable seed corporation decided in the summer of 2000 that it would "eliminate 2,000 varieties—or 25 percent of its total product line—as a cost-cutting measure."[62] The Rural Advancement Foundation explained the situation in a news release:

> Seminis, a subsidiary of the Mexican conglomerate Savia, controls nearly one-fifth of the worldwide fruit and vegetable seed market and is the source of approximately 40 percent of all vegetable seeds sold in the United States. The company built its seed empire by acquiring a dozen or so seed companies, most notably the garden and seed division of Asgrow, Petroseed and Royal Sluis. As a result of its buying binge, Seminis offerings grew to approximately 8,000 varieties in 60 species of fruits and vegetables. On 28 June 2000 Seminis announced that it would eliminate 2,000 varities—or 25 percent of its varieties, as part of a "global restructuring and optimization plan."
>
> No one knows for sure which varieties will be dropped from Seminis' commercial line, but the older, less-profitable open-pollinated varieties will be the first to go. Seed corporations favor hybrids because profit margins are greater, because gardeners and farmers can't save hybrid seed (thus encouraging repeat customers), and because newer varieties are more likely to be patented or protected by plant variety protection laws.[63]

The possibility that, if present trends continue, the majority of the world's farmers could become totally dependent for seed, fertilizer, pesticide, herbicide—even for the very structure of their increasingly simplified and ecologically vulnerable farming systems—on a small group of transnational corporations whose ultimate *raison d'être* is not food production but financial profit, and that the number of crop varieties available through these firms represents a drastically impoverished ecological potential, ought to give pause to planners everywhere.

It might well give them cold sweats, if the risks involved in the new processes of plant and animal bioengineering are factored into the equation. As the Environmental Defense Fund's Rebecca Goldburg explains:

> With traditional selective breeding, humans can cross one crop variety, say of potatoes, with another potato variety, and in some cases with related wild potatoes. But traditional breeders cannot add viral, insect or animal genes to potatoes—all of which *have* been added to plants via use of "recombinant DNA" and related genetic engineering techniques. . . . Scientists can now, at least in theory, take virtually any genetically encoded trait from one organism and add it to another, no matter how unrelated. . . .
>
> For those concerned about the environment, the prominent issue has been the possibility that scientists will inadvertently create "transgenic" organisms that wreak ecological or other types of havoc, particularly if [they] are intended for deliberate release into the environment.[64]

Exotic Species

Examples of the kind of mischief that can result when exotic organisms are released into environments where they have no natural ecological niche, and often no "control" in the form of local predators or competitor species, abound in history, particularly the story of European exploration and colonization. Crosby documents many disasters in the 1,000-year span of expansion of European civilization. For example, he describes the colonization of Porto Santo and Madeira by the Portuguese:

Madeira and Porto Santo were virgin in the purest sense of the word. They were uninhabited and bore no mark of human occupation, Paleolithic or Neolithic or post-Neolithic. The newcomers set to work to rationalize landscape, flora and fauna previously unaffected by anything but the blind forces of nature. Bartholomeu Perestrello, captain donatory of Porto Santo (and, incidentally, future father-in-law of Columbus), set loose on his island where the likes of such had never lived before a female rabbit and her offspring; she had given birth on the voyage from Europe. The rabbits reproduced at a villainous rate and "overspread the land, so that our men could sow nothing that was not destroyed by them."

The settlers took up arms against these rivals and killed great numbers, but in the absence of local predators and disease organisms adapted to these quadrupeds, the death rate continued to lag behind the birth rate. The humans were obliged to leave and go to Madeira, defeated in their attempt at colonization not by primeval nature but by their own ecological ignorance. . . . Europeans would make such mistakes over and over, setting off population explosions of burros in Fuerteventura in the Canaries, rats in Virginia in North America, and rabbits [again] in Australia.[65]

More recent examples include introduction of the European starling (*Sturnus vulgaris*) to New York City in the late nineteenth century, and of the infamous cane toad (*Bufo marinus*) to Queensland, Australia in the twentieth century.[66] The starling, introduced for esthetic reasons, reproduced at a prodigious rate, displacing and nearly wiping out the native North American bluebird and creating a nuisance with its huge noisy flocks. The cane toad, which is poisonous, was introduced to help control sugarcane beetle, but instead destroyed local Australian wildlife species and reproduced so rapidly as to become a virtual plague.

Plant species have exhibited similar behavior. For example, various crops transported by the Spanish to California and Peru:

They [the Spanish] did write about respectable plants that went wild and defied attempts to keep them out of cultivated fields, citing turnips, mustard, mint and camomile as among the worst offenders. Several of these "have overgrown the names of the valleys and imposed their own as in the case of Mint Valley on the seacoast, which was formerly called Rucma, and others." In Lima, endive and spinach grew taller than a man, and "a horse could not force his way through them."

The most expansionistic European weed in 16th century Peru was *trebol*, a clover or clovers that took over more of the cool, damp country than any other colonizing species, providing good forage but smothering crops as well. The former subjects of the Inca, who had abruptly found themselves with a new elite and a new God to support, now discovered themselves in competition with *trebol* for crop land. What was *trebol*? Most of it, in all likelihood, was white clover, which performed the same role of pioneer and *conquistador* in North America.[67]

Given such precedents, it is not difficult to imagine the results if transgenic organisms created in the laboratory to "improve" agriculture should escape farmers' control and establish themselves independently. Not only would such plants and animals—in such cases, perhaps more accurately called plant-animals or "plantimals"—have no natural controls in their immediate environments, they would have none in *any* environment, anywhere in the world.

The corporations that are developing such organisms, however, are eager to introduce them and much effort is being expended to overcome government constraints to their quick release. In the United States, the Department of Agriculture (USDA) is favorable to such introductions. Goldburg reports:

> In a highly controversial decision, the USDA in December 1994 allowed Asgrow Seed Company to sell squash genetically engineered to resist two plant viruses. The engineered squash will undoubtedly transfer its two acquired virus-resistance genes to wild squash (*Cucurbita pepo*), which is native to the southern U.S., where it is an agricultural weed. If the virus-resistance

genes spread, newly disease-resistant wild squash could become a hardier, more abundant weed.[68]

Another example of the unlooked-for consequences of introducing gene-altered organisms into the environment was noted in 1999:

> A popular new variety of corn plant that's been genetically modified to resist insect pests may also be taking a toll on the Monarch butterfly, new research suggests.
>
> The gene-altered corn, which exudes a poison fatal to corn-boring caterpillars, was introduced in 1996 and now accounts for more than one-quarter of the United States' corn crop—much of it in the path of the Monarch's annual migration.
>
> Pollen from the plants can blow onto nearby milkweed plants, the exclusive food upon which young Monarch larvae feed, and get eaten by the tiger-striped caterpillars.
>
> In laboratory studies conducted at Cornell University, the engineered pollen killed nearly half of those young before they transformed into the brilliant orange, black and white butterflies well known throughout North America.
>
> Several scientists yesterday expressed concern that if the new study's results are correct, then Monarchs—which already face ecological pressures but have so far managed to hold their own—may soon find themselves on the endangered species list. Other butterflies may also be at risk.[69]

The general public—not just those who own or operate farms—has an obvious stake in preventing reckless introduction of laboratory-created organisms that could threaten not only our food supply but also the overall environment in which we live. It follows logically that the news media ought to be monitoring closely and reporting developments in the seed trade and in breeding research.

Also crucial to the environmental well-being of the general public is the system of regulation that guarantees the safety and purity of the foods we eat. As mentioned in the discussion in chapter 3 of the effects of the GATT agreement, food safety regulations are now the subject of dispute between GATT signatories, some of whom regard such rules as indirect barriers to trade.

In the past, food safety standards were generally within the jurisdiction of national and sometimes provincial or state governments. Their strictness varied from country to country, with regulations in North America and Western Europe generally being more stringent than those of other nations. The new GATT regime, however, has imposed a single international standard: the Codex Alimentarius (Latin for "food code"), established and administered by a joint UN Food and Agriculture Organization (FAO) and World Health Organization (WHO) commission. Though Codex regulations are stricter than their national counterparts in many Third World and former East Bloc countries, many of the code's sections fall short of the standards of more technically advanced industrialized nations. Consumer groups in the north fear the Codex will be used to drag their national standards down, weakening protection against possibly dangerous food additives and other substances.

For example, in the United States synthetically produced growth hormones are routinely administered to dairy and beef cattle to stimulate milk production and/or weight gain. Low doses of antibiotic drugs are also regularly administered to perfectly healthy cows, not to treat disease but to prevent casual bacterial infections that might weaken an animal and slightly decrease its rate of milk production or weight gain. The possible effects of trace amounts of these hormones and antibiotics on people who ingest the resulting animal products has been the subject of controversy. Both Canada and the European Community (EC) have held that cattle fattened with hormones are unfit for human consumption and have banned imports of hormone-treated beef.[70] Research scientists have also warned that constant use of low doses of antibiotics could result in resistant strains of bacteria, dangerous to both animals and humans. Livestock producers in the United States, however, insist such practices cause no harm, and that the resulting food products are safe. They launched initiatives under the GATT to force introduction of U.S. cattle products to Canadian and European markets. The scientific evidence is as yet inconclusive, and the dispute continues. How it progresses, and its eventual outcome, should be closely followed by the news media.

Also of concern to informed observers is the widespread introduction of genetically altered foods to the shelves of grocery stores in North America, before adequate testing has been done to see if the new varieties could pose any health hazards to consumers. An advertisement posted in newspapers around the country in 2000 by the Turning Point Project highlighted the danger:

> You have the right to know if your baked potato contains bacteria genes . . . or if the tomato in your salad has genes of viruses spliced in. But at the very place where you encounter genetically engineered (GE) products— your local grocery store—there is silence.
>
> The Food and Drug Administration (FDA) and the biotechnology industry have prevented the labelling of GE foods, effectively subverting your right to know. *And so every day, millions of American infants, children and adults eat genetically engineered foods without their knowledge.*
>
> Are these unlabelled foods dangerous? *Nobody knows.*[71]

CHAPTER FIVE

Point of No Return:
A Clash of World Views

Let's recognize that we're in a real emergency;
scientists looking at whole systems long ago
recognized that. It's now so critical we should
seize the television and give everybody a
permaculture course.[1]

—Bill Mollison

Australia's Bill Mollison, founder of the Permaculture Institute and a harsh critic of the modern industrial agriculture system, is not known for beating around the bush. He puts it bluntly:

> Agriculture lost its way in the 1940s. Once it was there to produce food for people; now it's there to produce money for large interests. . . . Within 10 years certain very rigorous dies will be well and truly cast. . . . I hope that irresponsible systems collapse sooner rather than later. The longer they survive, the more damage they do. . . . there is a danger that traditional knowledge will die out and that there will be no one left to work the land in the sustainable sense.[2]

There are many who agree with Mollison, and within the specialized agricultural community of research scientists, policymakers, extension educators, and increasingly sophisticated industrial farm operators, perhaps an equal number who disagree. Whichever side the protagonists find themselves on, however, the fact is that—as the material discussed thus far should have made clear—a physical, economic, social, and intellectual watershed has been reached in agriculture. A similar crisis point has also been reached in most other harvest industries, such as forestry and fisheries.

It is difficult to exaggerate the importance of this clash of practices, and the worldviews they represent. Not only will its outcome directly—perhaps drastically—affect the stability and quality, not to mention the price, of everyone's food supply, it will also affect the long-term health of the environment and play a decisive part in defining the rural and urban social structure of future societies around the globe.

Two Camps, Two Systems

Essentially, two systems are in confrontation:

1. So-called *"sustainable"* agriculture. This features systems that are traditional, organic, local, labor-intensive, family-oriented, and based on diversity. Some of its leading modern exponents include Australia's Mollison, Japan's Masanobu Fukuoka, India's Vandana Shiva, and the father and son team of Robert and J.I. Rodale, founders of the Rodale Institute in the United States;
2. *Industrial agriculture.* This features systems that are non-traditional, inorganic, international, capital-intensive, market-oriented, and based on uniformity. Its leading exponents have included Green Revolution research scientists such as Norman Borlaug, staff of the Consultative Group for International Agricultural Research (CGIAR) centers, and the agriculture ministries of most of the industrialized countries, especially the U.S. Department of Agriculture (USDA).

Proponents of each camp have often been harshly critical of their opposites. Vandana Shiva, for example, attacks industrial farming:

The monoculture mind creates the monoculture crop. . . .
The destruction of diversity and creation of uniformity
simultaneously involve destruction of stability and the
creation of vulnerability . . . uses outside those that serve
the market are not perceived or taken into account.[3]

However, a rice farming specialist and Green Revolution
exponent at the FAO flatly insists:

If they adopted [traditional farming advocate
Masanobu] Fukuoka's methods in their entirety in
China, 600 million people would die.[4]

For more than two decades, the struggle at the research
level has more closely resembled the propaganda warfare of religious
zealots, or the political rhetoric of the Cold War, than it has a scientific discussion. Sustainable farming has been branded "muck and
magic,"[5] and industrial farming "death systems."[6] In some cases
even relatively disinterested pure research scientists, whose views
could nevertheless be interpreted as favoring one camp or the other,
have been openly muzzled or censored.

Such was the case with Health Canada researcher Dr. Shiv
Chopra, who in 1998 was barred by his government employer from
speaking at a public meeting on the subject of bioengineering of
foods. As the weekly paper *Capital City* reported:

[Chopra and others] claim public safety is being jeopar-
dized by a department whose puppet strings are being
controlled by large biotech corporations and their tasty
grants.

That was Dr. Chopra's message when he appeared
with fellow Health Canada scientist Dr. Margaret
Haydon on Canada AM, June 11. They stated that
Health Canada administrators were disregarding scientists' recommendations to withhold approval for agricultural drugs, thus endangering public safety.

When asked why there was pressure to approve
drugs so quickly, Chopra told the reporter "Well, what
do you think? Money. For multinational companies that
produce those things."

> Chopra received an official reprimand from Health
> Canada for appearing on Canada AM and is reticent to
> speak on the record now, for fear of further conse-
> quences.[7]

There have, however, been signs that at least some of the
rhetoric might eventually cool, and a few people of serious bent on
both sides might begin to listen to each other.

As long ago as the 1980s, for example, the announcement
by the U.S. Environmental Protection Agency (EPA) that agriculture
had become the largest nonpoint source[8] of water pollution in North
America spurred the Board on Agriculture of the U.S. National
Research Council to launch a study of traditional, organic farming
methods, comparing them to the industrial system. The study report,
published in 1989 under the title *Alternative Agriculture*, endorsed
many organic farming methods and shocked scientists and government administrators by charging that the U.S. system of farm subsidies and supports actually created "a variety of financial penalties that
farmers must overcome when adopting resource-conserving production practices."[9] In so many words, the report said that the deck had
been stacked in favor of industrial methods and against environmentally benign practices. Its conclusions were promptly and energeti-
cally attacked. But—to use an appropriate metaphor—a seed had
been planted. More such studies followed.

In 1992, the FAO magazine *Ceres* told its readers:

> With the growing stress on environmental limitations,
> "sustainable" and "integrated" are fast becoming catch-
> words of mainstream agricultural science, and a more
> objective look is being taken at what Green Revolution
> stalwarts once called muck and magic.
>
> Today, soil scientists and agronomists agree that a
> combination of organic matter, legumes and crop
> residues should be used to form the basis of sustainable
> cropping systems.[10]

On the other side of the divide, the well-known Dutch
Institute for Low-External Input Agriculture (ILEIA), a strong pro-
ponent of traditional, organic farming methods, notes in its most
recent mission statement:

Undoubtedly, there are some circumstances where external inputs are appropriate, while in other situations external inputs are too expensive or simply not available. However, the increasing number of success stories of farmers managing with low levels of pesticide, chemical fertilizer and local varieties can open the way to innovative alternatives which lie somewhere between the two options mentioned [sustainable and industrial farming systems].[11]

Both the Netherlands, with its long tradition of socially permissive, left-leaning governments, and New Zealand, site of one of the earliest national experiments in extreme right, neo-liberal economic planning, have adopted national "green plans" intended to completely remake their farm sectors along more traditional, ecologically appropriate lines.

Radical Change

The Netherlands' plan, approved in 1990 and now in its initial stages, calls for radical changes, including an assault on the country's elaborate system of dikes—made world famous in such legends and tales as "Hans Brinker and the Silver Skates." The dike system, centuries in the building, permitted the "low countries" to reclaim thousands of hectares of farmland from the sea. Twenty-seven percent of Holland's total land area consists of *polders*, or land reclaimed from as much as 20 meters below sea level. On that land, Dutch farmers have produced "some of the world's highest yields per hectare of meat, fruit, vegetables and flowers—among them the famous Dutch tulip and hyacinth bulbs."[12] Now, however, roughly 10 percent of the country's total farmland is to be taken out of production—and allowed to revert to wetlands, forest, and lakes. In some cases, dikes will be opened to permit water to flow back in. The change—prompted by a series of reports on pollution that predicted disaster for the Dutch environment and economy if changes weren't made—may well represent the greatest economic, social, and environmental event in modern Dutch history.[13]

The New Zealand plan is slightly less comprehensive in scale but also involves radical changes in farming systems. It was prompted by the realization that the 150-year-old system of sheep grazing in the country's rolling highlands—a highly profitable system that had made New Zealand wool synonymous with quality throughout the world—was not environmentally sustainable. Intensification of the grazing system in the past two decades had accelerated already serious problems of land degradation, soil erosion, and invasion by persistent weed species and pests, to the point where huge tracts had become barren and farmers were going out of business. Under the new plan, sheep grazing will be sharply curbed, in favor of such alternative land uses as grape growing, tourism, and cross-country skiing.[14]

Agriculture ministries in several countries, particularly in Western Europe, have been monitoring the Dutch and New Zealand experiments, and drafting possible green plans of their own. The fact that such complete reversals of long-standing—and highly profitable—systems have been adopted on a national scale in more than one country has sent shock waves through the international agricultural research community. As a spokesman for one nongovernmental organization put it in a conversation with this author: "The handwriting is on the corporate barn wall."

It should also be prominent in the major news media. Unfortunately, as the following chapters will show, this is not often the case.

PART II

THE IMPORTANCE ASSIGNED

TO AGRICULTURE BY THE

MAJOR NEWS MEDIA

AND JOURNALISM EDUCATORS

CHAPTER SIX

Perceptions of Decline:
Farm Journalism in North America[1]

*Around Halloween there are pumpkin farm
stories, around Thanksgiving there are often
turkey farm stories, and around Christmas,
Christmas tree farm stories.*[2]
—Max Armstrong, WGN,
U.S. Farm Report, Chicago

For the past 30 years, in newsrooms across the United States and
Canada, tales like that of Canadian journalist James Romahn have
repeated themselves:

> I was farm writer at the Kitchener-Waterloo *Record* for
> 20 years, took a buyout in 1974, and there was immedi-
> ately a significant reduction (in agricultural coverage).
> Mike Strathdee took over the beat, but it was added to
> his existing work as a high-tech industry reporter in the
> business section. He had no background in agriculture
> and food industry reporting, which was a major handicap.
>
> When I left, I was providing news and features cov-
> erage equal to any reporter covering any other beat, plus
> four columns per week, two beamed mainly at food
> industry audiences (farmers, processors, suppliers, etc.),
> one to consumers (the person pushing the shopping cart

through the local supermarket), and one to urban people as an education about farming. I also produced a regular weekly digest of news items in brief, in close to a column format, so you might say there were five columns per week.

All of the columns disappeared. What survives is a truncated version of the news briefs once per week.[3]

Romahn's successor, Strathdee, corroborates the tale:

My predecessor, Jim Romahn, was the dean of ag reporting in Canada, probably one of the last full-timers on Canadian dailies. When I asked to replace him I was told I could, as a part-time addition to everything else I'm doing. Started out as two days a week. Some weeks I'm now lucky to devote the best part of a day to the subject.[4]

Nor was the *Record* the only media outlet in the Kitchener-Waterloo municipality to cut back on farm coverage. As John Neil, of the Eastern Canadian Farm Writers Association (ECFWA), reports: "Similarly, David Imrie until recently was the farm broadcaster at CKCO-TV in Kitchener. They have just ended that position and Dave now covers general news."[5]

Readers' and listeners' reactions to this reduction in coverage were expressed in letters, phone calls, and other contacts. According to Romahn, most were negative:

I have experienced great frustration and anger from readers, cutting across a broad spectrum from farmers to consumers and food processors to retailers and researchers, mainly at local universities. The *Record* won two Michener Awards for meritorious public service in journalism based on my work, and about 80 national and regional awards for reporting and writing. It's all gone, with no realistic hope of ever being restored.

What's most frustrating to me is that in the 20 years I served as farm writer, three new layers of management were added. In successive rounds of severe downsizing, not one of those management positions has been eliminated.[6]

Neither reader reactions nor reporters' anger, however, seem to have done much to slow the reduction in farm writers' ranks. As already noted in chapter 1, the number of daily newspapers listing full-time agriculture or farm writers or editors on their news staffs has been in decline since the 1950s—dropping by well over 60 percent in the past 20 years alone in most of North America—while the number of AM and FM radio stations devoted to farm formats has also declined sharply.

The testimony is the same from all quarters:

- A 1988 survey by University of Idaho extension editor Clifton Anderson of journalists covering the U.S. presidential campaign found that "half of the 62 journalists interviewed insist it was unfortunate that farm policy issues remained on the back burner . . . While agreeing that agriculture's problems rate more discussion and political analysis, another group comprising one-fourth of the respondents said they were not sure it would be possible to heighten the awareness of national candidates and urban voters. The journalists who made up the remaining quarter of the survey sample said farm policy issues lack popular appeal and are unlikely to command much attention in future presidential campaigns."[7]

- Paul Queck, president of the American Agricultural Editors Association (AAEA), headquartered in Austin, Texas, says: "My observation is that the number of television and radio stations employing farm reporters is declining. That trend can be seen even among stations in agricultural communities."[8]

- "There is no question that over the past 40 years Canada's mainstream media have cut back drastically on personnel assigned to the agriculture front, and this during a period when resources for other types of coverage have expanded dramatically," writes Peter Hendry, a long-time farm editor in Canada who also served for 15 years as chief editor of the United Nations Food and Agriculture Organization's internationally distributed flagship publication, *Ceres* magazine. He continues:

In the early 1950s, when I was agriculture editor at the Winnipeg *Tribune*, I know that there were at least six, and perhaps more, counterpart positions on major dailies across western Canada. There would have been about the same number in southern Ontario. The 1992 *Gale Directory of Publications and Broadcast Media* listed only two daily farm editors for all of Canada—in St. John's, Nfld. (which eliminated the position in 1995) and Trois-Rivieres, Que., of all places. Perhaps half a dozen others across the country are listed as rural development editors, whatever that may imply.

In the 1950s, CBC Farm Radio had reporters based in every major region across the country, a service that has now totally disappeared. George Price was CBC's last full-time ag reporter in Ottawa.[9]

- A 1994 Agriculture and Agri-Food Canada Communications Branch Product Evaluation report described the effectiveness of the branch's efforts to place material in Canadian print and broadcast media. For its prerecorded program, "Update on Agriculture," the branch noted: "Of the 148 stations contacted . . . radio stations that actually use Update on Agriculture total 22 (14.86 percent). . . . When Update on Agriculture is used, it is usually aired early in the morning at 5:30 or 6:30 A.M."[10]
- A June 1993 panel discussion hosted by Agriculture and Agri-Food Canada and the Eastern Canada Farm Writers Association, entitled "Agri-food: the forgotten beat," solicited the views of Canada's leading farm writers (most of them represented specialist publications) on media coverage of farm stories. Noted Barry Wilson, Ottawa correspondent of the *Western Producer*: "The Parliamentary press gallery focuses each day on a few main stories determined by the political elite. Agri-food is not usually on this agenda. Agri-food is too complex and subtle to attract the gallery's attention on an ongoing basis . . . superficial treatment is the best that can be hoped for from mainstream media." *Canadian Press* Ottawa Bureau Chief Kirk Lapointe added, succinctly: "CP no longer has an agricultural reporter."[11]
- A 1995 list of "key agri-food media in Canada," compiled for internal use by the Communications Branch of Agriculture and Agri-Food Canada, warned that "because of

downsizing in the daily newspaper business, there are fewer reporters covering strictly agriculture. Accordingly, we are making efforts to target our messages, when applicable, to reporters covering other beats such as science, food, the environment, and business." The accompanying province-by-province breakdown of media outlets, intended for government public relations officers' use, included such comments as: "New Brunswick, *The Telegraph Journal*, like other dailies in the province the paper does not have an agricultural beat. Agricultural stories usually show up on the business pages. . . . Nova Scotia, *Farm Focus*, the only Atlantic-based farm newspaper . . . Newfoundland, *Western Star*, it's the only publication in the province that still focuses on agriculture on a regular basis . . . *Evening Telegram*, St. John's daily, recently discontinued the agri-food beat," etc.[12]

- Max Armstrong, farm broadcaster for WGN, U.S. Farm Report, Chicago, reports: "As an agriculture media person in a major U.S. city (one of the few), I can tell you that factual, unemotional ag economy stories are especially scarce . . . around Halloween there are pumpkin farm stories, around Thanksgiving there are often turkey farm stories, and around Christmas, Christmas tree farm stories. And that's all."[13]

- As noted in chapter 1, Project Censored, the media watchdog project originating in California, now has a Canadian branch, Project Censored Canada, a joint venture of the Canadian Association of Journalists and Simon Fraser University that "annually solicits and lists the most under-reported stories in the mainstream media."[14] Of the Canadian branch's 1995 "top 10 list of under-reported stories," numbers three, eight, and nine were farm stories, respectively dealing with the effect on agriculture of new GATT rules on plant patenting, the destructive effect on rural people of World Bank projects, and the environmental risks of fish farming.[15]

- Wayne F. Alda, a reporter at the South Bend, Indiana, *Tribune* and member of the U.S National Association of Agricultural Journalists (NAAJ), observes:

The withering in farm coverage in daily newspapers is a troubling trend that we in the NAAJ have been witnessing for some time. My observation only, but the trend seemed to get worse in the mid-1980s, about the same time local stations sought actors to anchor their news broadcasts. The NAAJ has long recognized that fewer and fewer of us are full-time farm writers. We've diminished in ranks (perhaps less than 100 nationwide), and many of us are now part-time farm writers with other beats to cover.

I guess I'm pretty representative of the multiple beat reporter. In addition to agriculture, I handle the environment beat, the weather beat when the ill winds blow, and I also tackle whatever science stories we pursue. So I could be writing about the Endangered Species Act, a Superfund investigation, a Cold War nerve agent storage depot, glaciers, a comet, and an extensive review of the particulars in the new farm bill (all issues, by the way, that I covered just last week). The week before I wrote stories about the dairy sector and the brain (actually a blood marker of the neurotransmitter serotonin that might be an indicator of suicidal tendencies). Oh yes, I cover the South Bend park department too. After I get done writing you, I will begin work on a story about some new discoveries of an old-line mosquito-borne disease called Cache Valley virus. Not to belabor the point, but there is only one full-time farm writer in the state of Indiana. He's Jeff Swiatek from the Indianapolis *Star*.[16]

- Tana Kappel, agricultural extension communications coordinator at Montana State University-Bozeman, reports: "In Montana, at the Great Falls *Tribune*, the ag editor was replaced with a farm and feature writer. The news hole for ag has dropped considerably and the focus is not ag news for farmers, but ag news for general readers. This change occurred about two years ago. The change was prompted by focus group surveys conducted by the *Tribune*.[17]
- In a 1993 *Columbia Journalism Review* article, Michael Balter described the results of a content analysis of reporting on negotiations leading to the GATT Agreement on Agriculture. He looked at coverage in the *New York Times*,

Washington Post, and Los Angeles *Times* from January 1, 1990 to December 15, 1992. Wrote Balter: "Most stories about the GATT talks have been relegated to the back pasture of the business pages . . . although the GATT talks obviously involve a multinational battle of interests, the print media's coverage has seldom ventured very far beyond the dry assumptions and calculations of U.S. trade officials. . . . The Los Angeles *Times*, for its part, despite its greater resources abroad, did not run a single detailed analysis in 1992 of the problems facing European farmers."[18]

- Observes Bob Davis, of *Iowa Farmer Today:* "The public as a whole does not appear to be well-informed about rural issues. Often times ag news does not receive the display it deserves. . . . When was the last time you stood in a supermarket checkout lane and read the covers of the magazines on display there? Did you see anything about agriculture? There you are, surrounded with agricultural products, and there is little information about those products."[19]

- A content analysis of Canadian daily newspapers produced for the Canadian Senate Special Committee on Mass Media showed that, as long ago as 1969, agriculture was already a neglected topic. The study listed 16 "areas of news interest" by subject matter and measured the percentage of items pertaining to each subject appearing in newspaper stories. At the top of the list were "human interest" (15.8 percent), "sports" (15.0 percent) and "politics and government" (12.7 percent). "Agriculture" was dead last, at 1.5 percent. A similar survey today would be almost certain to show an even lower figure.[20]

Further comparison of the daily newspaper staff listings in the *Editor & Publisher International Yearbook, 1995* helps to put the resulting situation in perspective. According to *E&P*, 574 out of a total 1,548 dailies in the United States employed a total of 690 full-time "amusement," "entertainment," or "film" editors or writers in 1995. In the same year, only 248 U.S. dailies employed full-time "agriculture" or "farm" writers or editors.[21] Thus, U.S. dailies devoted nearly three (2.78) times more of their full-time editorial manpower to covering amusement as they did to covering agriculture.

In Canada, the disparity was even greater. According to

E&P, 27 out of a total 107 Canadian dailies employed 34 full-time amusement, entertainment, or film writers or editors in 1995, while only eight daily papers employed full-time agriculture or farm writers or editors.[22] Thus, Canadian papers devoted more than three (3.37) times the full-time editorial resources to covering amusement as they did to following developments affecting the nation's food source.

The above trends showed no sign of reversing as the new millenium began. By 1999 (the *E&P Yearbook* is published in December each year) there were 1,489 dailies left in the United States, of which only 181 listed a farm or agriculture editor on staff—a 27 percent drop in the total number of such editors in a mere four years.[23] In Canada, the number of daily newspapers rose slightly, from 107 in 1995 to 109 in 1999, but the number of farm editors held steady at only eight.[24]

The Reasons Why

Looking at the evidence, the question is not whether agricultural coverage is declining, but why. The fact of decline is self-evident. But what logic dictates that amusements—or sports, fashion, music, or travel, all of which, as beats, involve more staff than farm news—should be considered more deserving of attention than the industry that supplies the food we eat? At first glance, such a position seems untenable, even absurd.

No systematic study of the question seems yet to have appeared in the professional literature. Nor is it within the scope of this thesis to attempt the kind of full-scale surveys—including in-depth examinations of advertising, marketing, and circulation functions and their connections with editorial content—that would be needed to find definitive answers. It is possible, however, to distinguish some potential lines of investigation.

Of marginal importance, but still a possible influence, is concentration of ownership and the decrease in the number of daily newspapers over the past few decades. In 1955, there were 1,765 daily newspapers in the United States.[25] In 1995, there were 1,548—a decline of 13 percent.[26] In Canada, there were 94 dailies in 1955,[27] a number that rose to 115 by 1975,[28] only to decline again to 107 in 1995—a drop of 7 percent.[29] It is possible that the smaller number of papers may account for a minor part of the more than 60 percent decline in farm coverage, but it is obviously not the leading cause.

In the U.S. farm state of Iowa, for example, there were 44 daily newspapers in 1955, of which 20—45 percent—had full-time farm editors.[30] In 1975, there were 42 dailies, a drop of 7 percent, but only 16 farm editors, a drop of 20 percent.[31] By 1995, there were 38 dailies in Iowa, a drop of 10 percent in 20 years—but only six farm editors, a drop of 63 percent over the same period.[32] The far faster rate at which the number of farm writers dropped, compared to the decrease in the number of papers, indicates some other factor was at play.

Anecdotal evidence suggests that the 1980s/1990s phenomenon of corporate "downsizing" has played a key role, as already indicated by the comment of the Communications Branch of Agriculture and Agri-Food Canada, cited earlier, that "because of downsizing in the daily newspaper business, there are fewer reporters covering strictly agriculture."[33] Managing editors, such as Kevin Cavanagh of *The Standard* in St. Catherines, Ont., echo this theme. Cavanagh's 40,000-circulation daily eliminated its farm beat "about 10 years ago," despite the paper's location in the heart of the Niagara Peninsula's orchard and vineyard country, where farming is a leading industry:

> It ceased to be a single beat and was combined with other subjects, because we just didn't have the number of bodies in the newsroom. At the time, it was combined with one of the lesser municipal beats. Today, farming could be part of the business beat, or sometimes under the environment.
>
> It was part of the trend toward having fewer and fewer autonomous beats, toward reporters becoming generalists. It's a reflection of the economics of the industry, the need to cut overhead, as well as of newspapers trying to be more things to more people. We hadn't done any reader surveys. It wasn't based on readership facts so much as the editors' idea of the number of rural readers being fewer, a judgement call. There was a need to cover more things with the same or fewer people, so we started to piggy-back.[34]

91

Peter Hendry also points to a changeover in some cities from broadsheet to tabloid format newspapers: "In the last 45 years, Calgary, Edmonton, Winnipeg, and Ottawa have all lost competitive broadsheet dailies, to be replaced—if at all—by tabs seemingly more interested in human than bovine mammalian devices."[35]

In Vancouver, the 161,000-circulation *Province* had been a broadsheet with a full-time farm writer in 1975, but subsequently dropped its farm beat and adopted a tabloid format. Says Managing Editor Michael Cooke:

> We turned tab about 12 years ago. I'm guessing, but I think the agriculture beat was dropped at about that time. We now have a lower mainland [urban area] focus. As you know, we're in trouble. We're struggling. It's a question of resources now. To be honest, I'd rather have a shopping mall reporter than a farm writer. Our readers are basically only interested in greed, energy, conflict, celebrities and shelter.[36]

The same kind of financial pressures and cost-cutting binges that have affected newspaper staffs have also had an impact on radio. As A.J. Blauer reports:

> Between 1988 and 1993, private AM stations lost approximately $231 million before taxes; private FM stations, though still in the black, have seen pre-tax profits dwindle from $27 million in 1988 to $14.5 million in 1993. Radio news, being the most costly to program, was the first to be pared. Newsrooms began shedding staff, reducing air time, and relying more and more heavily on imported news.
>
> In 1988, Tom Kark, news director of CJJR-FM and CKBD in Vancouver, saw his staff shrink from 13 people to two in just one day. Six years later, further deteriorations in the newsroom have him jokingly suggesting a "memorial for radio news."[37]

There seems little doubt that cost-cutting and layoffs—the "bottom line" mentality—have affected farm coverage. But decisions to cancel or consolidate beats aren't taken at random. They depend on the perceived priority of the subject. Newspaper amusement/entertainment beats, for example, are not being cut, nor are their writers being asked to double as business reporters or to cover city hall. As Michael Cooke put it, "celebrities" are regarded as too important to readers to risk neglect by dividing entertainment writers' attention. Rather, the beats that editors see as marginal suffer in "downsizing" exercises.

Why should agriculture be perceived as a marginal beat? The answer may lie in the precipitous decrease of the farm population, particularly since the Second World War, in advertisers' reactions to this demographic phenomenon, and to a fundamental misunderstanding on the part of editors and advertisers alike of the potential interests of urban readers. Although no direct investigation has been published in the literature, there are several indicators that the logic of such an explanation is strong.

Neither People Nor Money

"In 1850, agriculture made up a resounding 65 percent of all employment," writes demographer William Lazer. "Its share dropped continuously to only 3.3 million (2.7 percent) in 1990."[38] According to U.S. Bureau of Labor Statistics tables cited by Lazer, farmers and farm workers are among the "most rapidly declining" of growing and declining U.S. occupations.[39] Already at an all-time low, the number of farmers is expected to have fallen by more than 20 percent, and farm workers by more than 11 percent, between 1990 and 2005.[40]

Lazer offers further detail on overall U.S. farm population:

> At the turn of the 19th century, the United States was largely an agrarian society, and over 40 percent of the population lived on farms. In 1930, the farm population made up about 25 percent of our population. Then, a major transition occurred, for the farm population in 1990 accounted for less than two percent of our population.
>
> The continuing decline in both the total farm population and the percentage of the resident population they make up is striking. Former farmland continues to be paved over and transformed into suburban residential

> areas, towns, cities and shopping centres. There are
> fewer farm people living on fewer farms. . . . The average
> farm size more than doubled, from 213 acres in 1950 to
> 462 acres in 1987.[41]

As for farm incomes, the figures for Canada's wealthiest province, Ontario, are indicative. In 1992, the total average income of farm operators in Ontario was Can$31,663 (US$23,748), of which more than half—Can$19,100 (US$14,325)—came from *off-farm* income.[42] With an annual bottom line of only Can$12,563 (US$9,423) from farm sources, Ontario farmers are hardly among the province's high rollers. They are typical of farmers across the North American continent.

So marginal are farm incomes in the overall scheme of things, that such a basic measure of economic reality as the Consumer Price Index (CPI) does not even refer to rural groups. As Lazer notes, "CPIs are published for two population groups: all urban consumers (the general index) and urban wage earners and clerical workers."[43]

If, as Lazer observes in another chapter, "Markets are often defined as (1) people with (2) the money and (3) the willingness to buy,"[44] then the miniscule and still-shrinking farm population, with its low farm income averages, would seem to offer advertisers neither people (their willingness notwithstanding) nor money. Advertiser reaction to this situation, and its effect on editorial content, should not be difficult to deduce.

Modern news media live or die by the advertising dollar. As C. Edwin Baker notes, "at profit maximizing point, circulation revenue will predictably amount to a loss covered by profits on advertising."[45] Further:

> Advertisers want the readers most likely to buy their
> products. Depending on the medium's advertisers, a
> variety of types of media differentiation might result. For
> example, particular advertisers' preferences for particular
> audiences help explain profitable special-interest maga-
> zines. . . .
> Daily newspapers depend primarily on commercial
> advertising biased toward a very broad, relatively undif-
> ferentiated, middle to upper-middle market of people

with comparably large disposable incomes. As Michael Schudson puts it, "marketers keep their eyes on the main prize—pocketbooks, not persons." This can be a strong preference. . . .

Advertisers generally favor the paper with relatively more affluent readers, subsidizing these readers' preferred [editorial] content.[46]

So pervasive is advertiser influence over media content that it can result in severe, sometimes outright ridiculous self-censorship, particularly in television. As Baker explains:

Advertisers want more than editorial content supportive of their products. At least in broadcasting, they also want the surrounding content to promote a frame of mind that leaves readers or viewers most open to advertising messages. Thus, Du Pont "told the FCC that the corporation finds its commercials more effective on 'lighter, happier' programs." Bob Shanks, an ABC-TV vice president for programming, indicated that shows should "attract mass audiences without unduly offending these audiences or too deeply moving them emotionally (because, it is thought, this would) . . . interfere with their ability to receive, recall, and respond to the commercial messages."

This can be compared with the statement of a vice president of Coca-Cola: "It's a Coca-Cola corporate policy not to advertise on TV news because there's going to be some bad news in there and Coke is an upbeat, fun product."[47]

This attitude was perhaps carried to the extreme of irresponsible absurdity during the Gulf War, as Baker reports:

America's attack on Iraq in 1991 produced unusually explicit indications of advertisers' impact on news content. Beginning with the intensive coverage of the aerial bombing of Iraq, many companies decided to avoid advertising during war programming. One advertising executive explained about ads during war news: "I just think it's wasted money. . . . Commercials need to be seen in the right environment. A war is just not an upbeat environment." This advertising pullout produced serious financial problems at each network.

CBS executives admitted a willingness to allow advertisers' response to affect programming decisions. Even though its war specials all received higher ratings than other channels' entertainment shows, thereby indicating that viewers wanted this broadcasting, the low ad sales on war programming "made them economically unfeasible for the network." As a result, fewer prime-time war specials were shown. . . .

Since advertisers demand upbeat surroundings for their ads, CBS offered to tailor war specials "to provide better lead-ins for commercials . . . [and] to insert the commercials after segments that were specially produced with upbeat images or messages about the war, like patriotic views from the home front." A happy war brought to you by your sponsor![48]

Nor are such biases limited to the world of television:

Although the "buying mood" phenomenon applies mainly to broadcasting, the rough equivalent is the print media's introduction of surrounding editorial content provided not so much because readers desire or would pay for it or because editors judge that it ought to be printed, but because it focuses the attention of an advertiser-relevant group of readers and because its themes arguably increase these readers' interest in products the advertiser sells.[49]

Newspapers are so anxious to tailor themselves to obtain the sort of demographic profile sought by advertisers that they will deliberately jettison paying subscribers who don't fit the mold, and adjust content to attract those who do.

The *Times* (London) "twice increased circulation only to find advertisers were uninterested in reaching the students and lower income intellectuals who made up the bulk of the new readers. In order to stem this massive loss caused by having gained the wrong sort of readers (too poor, too working class), the *Times'* management in the early 1970s adopted a conscious policy of shedding part of the impressive 69 percent circulation gain it had obtained between 1965 and 1969 . . . the Toronto *Globe and Mail* "abolished its city desk and refocused coverage on business, international and national news," with the result that circulation dropped but ad revenue increased nine percent for the year.[50]

In such an advertiser-dominated climate, how are rural readers perceived? A particularly revealing clue was the cancellation of two popular "westerns" by network television executives, around the same time that newspaper farm editors' numbers first began to decline in earnest. As Baker notes:

The right audiences are crucial for television. "Gunsmoke" is probably the most frequently cited show canceled while still high in the ratings. The same apparently happened to "The Virginian." The shows' viewers were simply **too old and too rural to be worth much to advertisers.**[51]

More recently, the Des Moines, Iowa, *Register*, the leading daily in a so-called "corn-belt" state, made a significant choice:

To keep margins up to the corporate specs of its owner, Gannett, the *Register*, among other measures, abandoned its historic mission of circulating to all 99 Iowa counties and, instead, focused its marketing on the Des Moines metropolitan area.

"I wish we could concentrate on making money by putting out a good newspaper instead of trying to satisfy shareholders on a quarter-by-quarter basis," [Editor Geneva] Overholser said at the time.[52]

In short, the likelihood is very strong that the rural audience has simply been written off by all but farm equipment manufacturers or other such farm sector advertisers, who reach their audiences through specialist farm publications.[53] News executives in the mainstream print and broadcast media have reacted accordingly. Farm news has dropped to the bottom of the priority list.

Faulty Logic?

It is hard to fault the logic involved, provided the basic premises on which the apparent syllogism is based hold true: (a) readers' only value in our society is as potential advertising markets; (b) rural people are the only readers interested in farm news; (c) the rural audience is miniscule, in marketing terms virtually nonexistent, and what little there is of it has little money to spend; and (d) therefore, there is no reason to report farm news.

Premise (a), of course, ignores the basic need in a functioning democracy for an informed citizen base. As William Blankenburg has explained, "the trouble with expelled subscribers, whether they meet marketing standards or not, is that they are citizens."[54] To ignore rural readers, either in terms of circulation or editorial content, is a form of civic irresponsibility.

More practically important perhaps, in today's pro-business, anti-altruistic public climate, is the likelihood that premise (b) could also be mistaken. It appears based on the assumption, apparently unsupported by evidence, that *urban* readers—who are the majority and do have money to spend—have no interest in agriculture or rural affairs, and could not be induced to develop any.

The assumption of this text is that this premise is flawed, just the way the reasoning behind CBS's decision to report only "happy war" news was flawed.

James Rohman has already been quoted in witness to the "frustration and anger" of readers—many of them urban dwellers—faced with his paper's decision to scale back its farm reporting. Peter Hendry and other farm writers also stress the need "to explain agriculture to city dwellers, in terms of how it affects them."[55] As Hugh Owen of the University of Guelph comments:

> There is still a great interest in farming on the part of the consuming public, simply because they know farming has changed. . . . I know this from my participation on CBC's Radio Noon "Farm Panel" in Montreal, weekly successor

98

to the now-defunct daily "Farm and Food Show." Styled after the Gzowski panels on "Morningside," the host, two farmers and myself cover a range of topics, from issues (rBST) to news (beef marketing board starts electronic auction). Despite the fact that "hard core" farm news (no misrepresentation implied) is not regarded as appealing by the radio people in Montreal, I have received a lot of good feedback from urban listeners; they like to hear about what's going on down on the farm, even if it is quota prices and piglet tails. They wouldn't likely read a farm newspaper to get that news, but they don't mind hearing about it in order to give themselves a sense of what's going on.[56]

More convincing proof of the potential marketability of media vehicles that interpret rural material to urban audiences can be seen in the astonishing success of Canada's *Harrowsmith* magazine— which went from zero to 135,000 readers in less than three years from startup in the late 1970s—as well as that of such U.S. publications as *Mother Earth News, Blair & Ketchum's Country Journal,* and the U.S. edition of *Harrowsmith,* launched in the mid-1980s. Throughout its early history the Canadian *Harrowsmith,* which carried a considerable amount of "hard" farm reporting (e.g., its 1982 multi-author, in-depth examination of farm marketing boards, later reprinted by Agriculture Canada[57]) as well as extensive agricultural "how-to" material, maintained an urban readership of up to 60 percent. Even today, in its rather watered-down form (the magazine's format was changed when it came under new ownership) it still maintains a 50/50 urban/rural split.[58] Critics have charged that the original *Harrowsmith* was an aberration, a sort of literary wish book for urban "wannabees" who dreamed of an idyllic life in the country much as Britain's Elizabethan nobility dreamed of mythical nymphs and shepherds dancing on the green. Whether this was, in fact, the motivation behind the magazine's urban subscriber base is arguable. But it does not change the fact that—regardless of their motivation—urban readers *were* willing to pay good money for the original *Harrowsmith's* detailed, critical news about farming and rural life.

Even if urban audiences are not now favorably disposed to receiving farm news, it is not unreasonable to expect that they *could* be: that an audience might be created if readers and listeners were helped to realize agriculture's importance to their own lives and livelihoods—an importance I have tried to demonstrate in previous chapters.

It is unlikely that much of a following will be built, however, by relying on untutored general assignment reporters, rather than journalists with some background in agriculture, to interpret the industry to non-farm audiences. As the survey by Ann Reisner and Gerry Walter, mentioned earlier, revealed:

> Because non-governmental sources of agricultural news generally are outside the news-gathering "net" and because general newspaper reporters lack a ready "frame" for casting agricultural phenomena as news, agricultural news is a relatively minor component of the editorial content of papers (and other media) that have primarily urban audiences. These factors also assure that most news coverage, including any agricultural news that does appear in the paper, will be event-based and conflictual. The same norms lead newspaper journalists to portray non-news aspects of agriculture in mythic terms that have broad public appeal. Gans pointed out that feature stories about agriculture tend to reflect urban journalists' (and American society's) agrarian notions of the moral superiority of agriculture and to project overly-nostalgic views of farming and rural life.[59]

William B. Ward, in his landmark textbook *Reporting Agriculture* (perhaps the only such "how-to" text published), gives a still less flattering description (strongly reminiscent of the urban/rural cultural conflict described in chapter 2 by Williams):

> More fundamental than so-called advertising pressures is the fact that daily newspapers are city products and that inevitably they have shared the urban philosophy, prejudices and blind spots. In other words, agriculture has to deal with the agelong chasm between people in the cities and people who work the soil. A two-way bridge is needed over that chasm. The urban press, with a few notable exceptions, has contributed little to the building

of a bridge and is still handicapped enormously by its own background.

Basically, most executives of press associations and newspapers are not farm-minded. They do not think that the rather difficult, good farm news is worth the expense and spadework to obtain. Furthermore, some of them say that nothing ever happens to farmers worth much space when there is news available and easy to get at low cost from the police stations and courts. When a New England man writes that "the city slickers who comprise the staffs of certain important papers" are "interested in agricultural news only as it affects the price of milk," and that "a two-headed calf born in Skewhegan is more important" than are significant farm developments, he is hitting not merely at habits of sensationalism and super-ficiality, but at the deeper cause—at the prevalent city notion that agriculture simply can't make real news.[60]

Typical of the ignorance of city reporters when covering rural stories is the attitude toward firearms. Rural people grow up in an atmosphere that accepts firearms use as normal, sometimes necessary, and the average farm person has a fairly thorough grasp of firearm terminology.

Country people wince every time they read a story in a newspaper that talks about a "bullet from a shotgun."[61] Shotguns don't fire bullets, rifles do. Only a "city boy (or girl)" would make such a verbal blunder.

Rural people have even less patience with government bureaucrats—and the urban-based reporters who quote them—when they launch upon crusades designed to solve various urban crises, but that have unforeseen consequences in the countryside. Canada, for example, recently embarked on a zealous project to require new licensing and purchase requirements for gun buyers. The new laws were a reaction to a spate of shootings in major cities, involving inner-city people, some of them drug addicts, who used pistols to commit murders.

The framers of the new laws failed to consider their effects upon rural people, such as the Inuit residents of Canada's newest province, Nunavut. Under the new laws, every gun owner must pay a variety of licensing and transfer fees, which in the case of some

Inuit families may total more than the family's total monthly income.

The Inuit culture is a hunter-gatherer culture, developed over thousands of years in the harsh Arctic environment. Hunting seal and caribou is integral to that culture, and to Inuit survival. "Unlike in southern Canada, we can't domesticate the animals that we rely on so we need rifles in order to harvest our food that we put on the table," explained Nunavut Premier Paul Okalik. "This law would make it very difficult for some of our residents to do that."[62]

No one seems to have thought of this when the new laws were drawn up by ignorant, urban-based legislators.

The solution to this problem, and perhaps to the overall problem of the "invisible farm" itself, may be a question of education.

Education and Training

As noted in chapter 1, orientation in agriculture is hardly a priority among the journalism faculties of U.S. and Canadian universities and colleges.[63] Of the 510 U.S. university schools of journalism listed in the membership directory of the Association for Education in Journalism and Mass Communication (AEJMC), only seven had any courses in agricultural journalism.[64] Though not listed in the AEJMC directory, Colorado State University also offers some courses in farm journalism, for a total of eight U.S. universities.[65] Typical of the attitude of U.S. journalism schools is that of East Texas State University: "We do not have a program in agricultural journalism at East Texas State University in Commerce. At one time we had a joint major with agriculture, but that program was dropped for lack of interest and because the department did not feel it could support such a program."[66] In Canada, where nine universities and ten community colleges offered courses in journalism in 1995, only one postsecondary program included any instruction in farm journalism.[67] Agricultural Communications courses and programs, similar in some respects to programs in agricultural journalism, are offered by a number of universities with substantial agricultural science faculties. A 1990 survey by Ann Reisner of the Department of Agriculture of the University of Illinois at Urbana-Champaign located 16 U.S. institutions offering such courses.[68] Canada's leading agriculture school, the University of Guelph, offers a single course in Agricultural

Communications as part of its Rural Extension Studies program.[69] Although this particular course does have a journalistic orientation, most agricultural communications programs are not oriented toward preparing students for careers in journalism as such. They are more often geared toward producing farm extension personnel, ag-industry public relations representatives or advertising specialists, or filling "jobs with university information offices, environmental and natural resource agencies, research institutions, private industry and special interest groups."[70] Graduates of such programs who enter journalism generally find work with specialized farm publications, rather than the mainstream press or broadcast media.

Ag-communications programs are frequently regarded with a jaundiced eye by the universities offering them, as Reisner reports:

> Several communications faculty members said that they believe agricultural communications course work wasted university resources. These faculty members generally gave the impression, and some stated specifically, that communications schools teach the skills necessary to cover a wide variety of topics, agriculture among them. Specific programs for agricultural communications students were unnecessary. The reaction from agricultural faculty members was, not surprisingly, different. While agricultural faculty felt that agricultural communications courses were important, several mentioned that limited university resources or higher level administrators' reluctance to invest in agricultural communications severely limited their program's growth potential.[71]

Given such views, in an era of "downsizing" and budget cutbacks, the future of agricultural communications in North America—with 16 programs versus eight, twice as popular as agricultural journalism *proprement dit*—seems far from secure.

This does not bode well for focusing mainstream media attention on agriculture.

CHAPTER SEVEN

Chaos Unobserved:
From *Kolkhozniki* to *Fermeri* in the East

*Until recently, historians of Russia did not
study the peasant majority of the population. . . .
Not a single major English-language
investigation of the topic appeared until 1968.*[1]
—Esther Kingston-Mann

Like the history of Russia, the story of agriculture in the former
Soviet Union and its satellite states is largely a record of suffering,
tragedy, and massive waste, not only of the lives of rural people but
also of the land itself. Since the fall of the Soviet system, some of the
waste and suffering has begun to ease, particularly in such ex-satel-
lite European countries as Hungary, Poland, and the Czech
Republic. But in Russia and the new nations that make up the still-
shaky Commonwealth of Independent States (CIS), the rural picture is one of chaos and confusion—and continuing suffering and
waste.

Just as historians and academics for years ignored rural
Russia as a subject of study (the first international conference *ever* to
focus on the peasantry of European Russia met in 1986[2]), so the
news media today are overlooking many of the momentous changes
going on in the agricultural hinterlands of the former Union of
Soviet Socialist Republics (U.S.S.R.) and its now-defunct empire.

Although the effects are the same, the reasons for this
neglect are different than those causing the decline in rural report-

105

ing in the West. Also unlike the West, where the decline in agricultural reporting has coincided with a drop in rural populations, the neglect of agriculture in the Russian press today coincides with a recent movement of population *back into* the country, similar to that seen in North America during the Great Depression.

In order to understand what is going on in the region, some historical background is necessary.

Enslaving Serfs, Killing Kulaks

The medieval institution of peasant serfdom, abolished in most of Western Europe by the late sixteenth century,[3] continued to function in Russia until the middle of the nineteenth century—and the life of the Russian serf was far harsher than that of his other earlier French or English counterparts.

In Russia, the serf was almost literally the landowner's slave. Unlike the English who, as mentioned previously, saw the yeoman freeholder—which any serf could become by purchasing his freedom—as the backbone of the realm, "staunch and valiant,"[4]—the Russians saw not only serfs, but all rural dwellers other than the aristocracy, as basically pernicious "pre-humans."[5] Nor did official emancipation of the serfs in 1861 change this general prejudice:

> Raising the peasantry out of the degradation of serfdom and the barbarisms of village life was a common theme for mid-19th century proponents of emancipation; in later decades, revolutionary "uplift" blended with elements of racism in the Marxist G.V. Plekhanov's description of the Russian peasants as "Chinese, barbarian-tillers of the soil, cruel and merciless, beasts of burden whose lives provided no opportunities for the luxury of thought."
>
> The powerful bonds that linked members of the peasant "herd" (the livestock image was used both by the Tsarist Minister of Agriculture A.S. Ermolov and by the Marxist Leon Trotsky) bred an ignorant hostility to outsiders, which erupted in pogroms against Jews and the lynching of doctors suspected of being the carriers of the diseases they came to cure. Loyalties to family and

community were described as "chains" and "fetters" which prevented peasants from exploring the opportunities available to them in the world outside. In order to destroy the ties between peasant children and their primitive parents and neighbors, a leading progressive educator of the 1880s demanded that peasant children be physically removed from their homes.[6]

If even the would-be benefactors of rural people had such a negative view of them, it is little wonder that the lives of many farm families after the emancipation of the serfs saw only marginal, if any, improvement. The Russian economic system of the late nineteenth century was, in fact, set up not to improve the lot of the freed serfs, but to channel them into a new form of semi-slavery as migrant laborers and city factory workers. Jeffrey Burds describes the arrangement:

> Peasant families could not make a living off their allotments, and were compelled therefore to rent additional land or work as wage laborers in the agricultural and non-agricultural sectors. In the Central Agricultural Region, such terms were imposed deliberately, so as to ensure a large, inexpensive pool of wage laborers for demesne agriculture . . . the typical redemption price for land in central Non-Black-Earth districts was one to two times its actual value among former state peasants, and two to two and a half times for former manorial peasants. In this way, the Russian fiscal system operated on the presumption that peasants in the non-chernozem areas would turn to sources outside agriculture to supplement their earnings and fulfill state obligations. Three other factors—a steady increase in rural tax burdens throughout most of the last half of the 19th century, the peasant's growing reliance on the market as a purveyor of needed items, and a rapid population growth—gave further impetus to peasant migration for supplemental earnings. . . .
>
> To meet the excessive obligations of state tax and redemption dues, peasants in the Central Industrial Region were compelled to depart for earnings outside their native villages.[7]

Exploited by middlemen and harrassed by officials, the *otkhodniki* (departed ones) migrated from region to region, village to village, as well as to the newly industrializing cities, finding work wherever they could, while their wives, children, and elderly parents stayed at home and did their best to farm the hard-pressed family plot. In many respects, they led lives similar to those of the exploited Mexican migrant farm laborers in the United States, who flocked to the union banner of Cesar Chavez in the 1960s. Most estimates agree that by the mid-1890s the total number of *otkhodniki* numbered:

> near two million for the nine provinces of the Central Industrial Region, more than six million for the whole of European Russia. This corresponded in the Central Industrial Region to more than 14 percent of the total rural population: more than a third of all adult peasant males, at least one member of every peasant household, were involved in some form of work that took them away from their villages.[8]

In such circumstances, only the cleverest and hardest-working of peasants—the so-called kulaks—managed despite the system to better their situations, gradually turning their small plots into medium-sized holdings on which they could make a modest profit. The kulaks—later unfairly caricatured by the Bolsheviks as rapacious village moneylenders—were the most knowledgeable farmers, those who best understood the land and its crops and could maximize its productivity.

They were helped—at least, those who could read—by the only popular farm publication of the day: *Sel'skii vestnik* (the *Village Herald*), a newspaper published by the Tsarist government for the Russian peasantry from 1881 to 1917.[9] At its peak in 1905, *Sel'skii vestnik* had a circulation of 130,000, scattered throughout the rural regions of Russian-held Poland, European Russia, Siberia, the Caucasus, and Central Asia.[10] It offered news and advice on up-to-date farming methods—but disappeared permanently, with no replacement, after the Bolshevik revolution.

As for the mainstream press in Tsarist times, its coverage of rural issues suffered from the patronizing attitude shared by many intellectuals and reformers of the day, as well as from constant, often severe government censorship. An example was press coverage of the so-called "vodka protests" of 1859, a peasant temperance movement that had revolutionary overtones. The protests, which included riots of such violence that the army had to be called out to aid local police, spread to eight provinces and saw hundreds killed and injured, or arrested and sent into exile in Siberia. At the height of the riots, "all newspaper discussion of vodka protests ceased, presumably at the instructions of the government censors."[11]

The anti-vodka movement, which had much to do with peasants' desire to end the institution of serfdom (sales of grain for vodka production were used by serfs to pay for their freedom), was viewed by many reformers as a simple anti-drinking crusade:

> The first attempts to explain the liquor protests were made by the progressive-minded observers from the gentry class who reported on it to national newspapers and journals in the early months of 1859. Almost all the early reports portrayed the protests as a temperance movement, and by March several newspapers were running regular columns headed, "On the spread of sobriety."[12]

One correspondent took a different view, seeing the protests as essentially economic—which they may well have been—but he too ran into the government's censorship apparatus:

> The social and literary critic N.A. Dobroliubov wrote the first detailed analysis of the liquor protests in *Sovremennik* in September 1859, (though he had had to wait for two months, and delete one-third of his original draft before the censors would pass his article).[13]

The situation did not improve, for peasants or agricultural news coverage, under communism.

As G.V. Plekhanov's comments, mentioned earlier, indicate, the Bolsheviks had as low an opinion of Russia's peasantry as their Tsarist predecessors. In fact, they had a special animosity toward the kulaks, whom they regarded as part of the hated "capitalist oppressor" class. While the peasants themselves—an exploited and oppressed group if one ever existed—hoped for better things under communism, their hopes were dashed, particularly once the dictatorship of Josef Stalin was established. As Orlando Figes explains:

> In 1917–1918, the Russian peasantry sought an agricultural system based upon the smallholding family labor farm. The peasant mandate, that formed the basis of the Decree on Land, called for the abolition of hired labor and the allocation of all the farmland to peasant households on a family labor norm. Wage labor, which facilitated large-scale capitalist and gentry farming, could play no part in the peasant family farm system.[14]

The reality under Stalin's new dispensation, however, was the diametric opposite. The ignorance of rural realities displayed by the Communists, and the scale of brutality employed in enforcing the new system, was staggering. Two generations later, people still remembered it with horror, as journalist Scott Shane found when he visited a village south of Moscow that had borne the brunt of Stalin's onslaught. Shane's visit came during the Gorbachev regime, when people could at last speak openly of the past to foreign newsreporters:

> People in Plosko-Kuzminka remembered well how Russia had been hammered into the first country of collectivized agriculture, a place of factory-farms where laborers traded a farmer's varied dawn-to-dusk labor for shift work at a single task. Plosko-Kuzminka demonstrated how Stalin had supplanted the myriad individual decisions that make up a market with a rigid hierarchy and top-down control, effectively locking the peasantry into a new form of the serfdom they had escaped 70 years earlier. . . .
>
> The unhealing wound of Plosko-Kuzminka was not the Great Patriotic War, World War II. Though it had taken the lives of many villagers, that war had been a just cause. No, the villagers interrupted one another telling the only foreigner they had ever seen about de-kulakization, the "liquidation of the kulaks as a class" at the beginning of the 1930s. The kulaks were the "rich peasants" who had managed to make a go of farming, build up a small herd of livestock, perhaps hire a laborer or

two. Naturally, they resisted the hardest when the Soviet regime ordered all peasants onto collective farms. In Stalin's determination to break the resistance, millions of kulaks were arrested and executed or sent into exile, along with countless ordinary peasants caught in the frenzy of score-settling and denunciation. "It was tearing everything down and building nothing in its place," said Ivan Fyodorovich.[15]

Some historians estimate that as many as 10 million rural people were murdered or starved to death—more than the victims of Hitler's Holocaust. Their deaths, unremarked in the Soviet press of the time, were remembered only by their friends and neighbors, who years later "told of families broken up, neighbors hauled away, houses burned, children going hungry as agriculture was devastated by the war on the kulaks."[16] What that stupidly misguided war was doing, via a deliberately planned, relentless process, was destroying the knowledge-base of Russian agriculture, the collective memory of those who had been the nation's most able farmers. The French have a word for it: *decervellezation*, "de-braining." As one survivor told Shane, "the kulaks, they were the *rabotyagi* (the hard workers)" who knew how to farm better than anyone else. And the Soviet Union, for the next 60 years, paid the price for their destruction. Agriculture in the new Utopia never recovered. Its crippled life was recorded, as Shane accurately notes, in:

> the long, sad story of the failure of the local *kolkhoz*, or collective farm, with its succession of grand plans from Moscow and its succession of hapless chairmen. Officials had tried the old Bolshevik name magic to will the kolkhoz to a "glorious destiny." What started as the First of May Collective Farm faltered and was recast as the Lenin Collective Farm. Later it was reorganized and renamed the Victory Collective Farm. In defeat, Victory was reorganized yet again to become the 50th Anniversary of the October Revolution Collective Farm. Finally Gorbachev ended the fashion for sloganeering names, and the local kolkhoz became the State Variety Testing Station. But none of the succession of names had conjured crops from the rich black earth.[17]

The trouble was, very few of Stalin's "outdoor factory workers" knew, or cared, what they were doing. As former *sovkhoz* (state farm) worker Ivan Petrovich Gresev put it in a recent interview: "When I was director of the state farm, I would give an order to two deputies, who would assign the job to four to eight agronomists. They would explain the procedure to their assistants, who would order the tractor drivers to carry it out. But the tractor drivers were always drunk!"[18]

And no wonder.

Praising Folly

Meanwhile, the Soviet press's reporting on agriculture followed the same model as its other editorial productions, expending millions of liters of ink in praise of folly. The praise, couched in the combination of jargon and inflated statistics that provided the material for so much western satire, continued almost unchanged from Stalin's reign through that of Nikita Khrushchev in the 1960s. As John Murray reports:

> Khrushchev himself as party first secretary tirelessly travelled the country giving many speeches, all covered by the press, which supported uncritically his initiatives in government administration and agriculture, including most notoriously the unsuccessful attempt to promote the widespread cultivation of maize in unsuitable climatic conditions. Just as the Stakhanovite movement had generated hero-workers in the press at any cost, so the desire to comply with the expectations of campaigns launched by Khrushchev meant that the figures and statistics which were obtained and paraded in the press were often at variance with common economic sense.[19]

Breakdowns, poor harvests, policy errors, and even bad weather had been officially ignored under Stalin, and "the practice of not reporting disasters and calamities continued unchanged under Khrushchev."[20] As one government functionary explained: "According to the newspapers we had no airplane crashes, our ships never sank, there were never any explosions in our mines, cars never ran over pedestrians, avalanches never fell on mountain towns and towns were never threatened by flooding."[21]

So low was the credibility of the Soviet press that ordinary Russian people actually assumed that if something was reported, the opposite was probably true. Commented the same functionary: "If you write that two people died, then everyone will think there were a lot more casualties. . . . If we wrote that a film was rubbish, there would be queues outside the cinema to see it, and vice versa."[22]

Typical journalistic devices of the era were the page-one *zametka* feature article, which reported—always in glowingly idealistic terms—on enterprises deemed worthy of emulation, and the *fotoreportazh*, also played on page one, which included a feature photo and praised individual people as role-models.[23] The requirements for such stories were clear: "For the journalist and rural correspondent it is always important to take note of and display the framework of things that are new and communist, to provide coverage of the patriotic deeds of the toilers and to provide information on the most important achievements of the Soviet people."[24]

Kolkhoz and *sovkhoz* workers were frequent subjects of *fotoreportazh* articles, which:

> typically contained biographical details of the heroic subject's life or work, often embellished a link between these details and the attainment of an economic goal, often expressed with the aid of one of the stock of current affairs cliches and large figures or statistics. The heroes of the *fotoreportazh* were as a rule shown to possess ideological motivation, usually expressed in cliched sociopolitical language.
>
> Thus, for instance, the "Cotton of Uzbekistan comes on stream" *fotoreportazh* (9 October 1979) contains *zametka*-like raw statistical information—the attainment of the goal—as well as figurative cliches applied to the work undertaken and to those who have undertaken the work.[25]

The language of such articles was as stilted and prescribed as that of a medieval morality play or the Japanese Noh theater. Individuals, for example, were referred to in ideological terms: "a progressive machine operator," or "progressive workers of socialist competition," or "creators of mighty tractors."[26] The "unchanging, fundamentally propagandistic philosophical intent" of all coverage was a constant that did not cease, even after Gorbachev's *perestroika* (1985–1991) had partially loosened the rules.

There is no doubt that the changes under Gorbachev were major, and they were felt in every walk of Soviet life—including agriculture and news coverage of agriculture. By the time Gorbachev came to power, it was impossible for the Kremlin's rulers not to see that the Soviet Union had become dependent on grain imported from abroad—chiefly from Canada and the United States—and officials of the new administration realized the old collective farm model had to go. They wanted to replace it with something closer to the western model, something, in fact, like the system originally envisioned in the aborted Decree on Land of 1918, mentioned above.[27] The government decided to encourage former state or collective farm workers to lease plots from the state farms and operate them as private enterprises—thus becoming *fermeri*, or "private farmers" (the term was borrowed from the English word "farmer").[28]

But even the Gorbachev regime was still part of a centrally-planned, communist system, and changes came slowly, those in the press included. Essentially, they were half-measures. For example, compared to the formal, rigid rules of writing in, say, 1979, "the language of the 1987 *fotoreportazh* is marked by the presence of colloquial lexicon, first person narration, parenthetical phrases, and particles, all lingusitic features characteristic of informal Russian and absent from the 1979 *fotoreportazh*,"— and all signs of the thaw in conditions under Gorbachev. But regardless of the more relaxed tone, they still uncritically reflected "the political mood of the leadership of the time, especially in the area of agriculture. Two of the headlines, 'Family farm (17 March 1987) and 'Family lease farm' (2 December 1987), show the government's backing for the family farm unit."

As Murray explains:

> The innocent-looking change in headlines masked a much deeper shift in policy. The government initiative to encourage peasants to lease land from the collectivized farms was an indirect challenge to the wisdom of the collectivization campaign of the 1920s and 1930s. Public condemnation at this [perestroika] period of both the

means by which agriculture was collectivized and the
failures of the collectivized system to meet the country's
needs was not possible, especially in the government's
own daily. Nevertheless it was impossible to avoid noticing the new emphasis in the press on the non-ideologi-
cally stimulated, small collective of an emotionally close-
knit and preferably large family that might include
grandparents and grandchildren. Thus the chatty and
familial style.[29]

Regardless of the comparative rigidity or flexibility of individual regimes, Soviet agricultural news coverage, like Soviet news
coverage in general, was never of high quality or very reliable.
Neither the workers on the state and collective farms nor the general public got much useful information from it.

Confusion's Reign

Nikita Khrushchev once told an audience: "You ask someone: do
you understand agriculture, and he tells you: what do you mean do
I understand, I've eaten potatoes! You see, once he's eaten potatoes,
that means he thinks he's an expert."[30] Now that the Soviet empire
has fallen, it seems, everyone in Russia and the CIS has eaten potatoes—yet no one has. Everyone knows the old way didn't work, but
no one knows what to replace it with. Farm workers and government officials alike are scrambling for answers, trying, in
Solzhenitsyn's famous phrase, to climb out "from under the rub-
ble." And there is very little help available in the news media.

The changeover has, in fact, been difficult throughout the
entire former East Bloc—even in countries like Poland, where a
large and stubbornly resistant peasant class, bolstered by the
Church, managed to keep 75 percent of farmland in private hands
and survive the Communist era more or less intact,[31] or in tiny
Albania, where the rigid Stalinist policies of dictator Enver Hoxha
had been so extreme that they prevented the country from developing a modern economy, and assured that nearly two-thirds of the
population remained rural.[32]

After the fall of communism, which had waged a relentless
(even if losing) war on Poland's peasants, preventing them from

obtaining credit or modernizing their operations, the country's private farmers at last were free of their longtime enemy, but also without friends, without modern farm equipment and machinery, and without capital. And there were too many farms: more than three million in 1991, compared to only 2.4 million in the geographically much larger United States. Roughly half were too small—between 1.5 and five hectares—to be economically viable.[33]

The new, non-communist government, faced with the harsh loan terms of the International Monetary Fund (IMF) and other international lenders, had few funds at its disposal to help the farm sector modernize. Officials estimated that some 1.5 million privately owned Polish farms would go out of business by the year 2000, thus accomplishing what communism had failed to do.[34]

In Albania, where forced collectivization had created a buildup of resentment and land-hunger, the fall of communism led to heavy public pressure for land reform. The result was a case of "privatization" gone wild:

> Basic statistics tell the tale: in 1988, the average state cooperative farm boasted 1,200 inefficient hectares, earning low but reliable prices from the institutions of central government. Land reform has subdivided those estates into 400,000 private plots, each averaging 1.6 ha—but without the tools, fertilizers and markets to make them viable. In the first season after privatization, wheat production fell from a 1980s average of 800,000 tons to 400,000 tons.[35]

In 1994, more than 2 million of Albania's total population of 3.5 million were dependent on food relief donated by the European Community (EC).[36]

Bad as the post-communist experience has been in the ex-satellites, however, it has been even worse in the CIS states themselves. A host of troubles has beset the farm sector. Where once an inefficient, centralized system existed to assure at least a minimum flow of fuel, seed, fertilizer, and spare tractor parts throughout the empire, as well as the slow, often wasteful, but at least functioning shipment of farm produce to market, a sudden vacuum ensued. The newly independent states had not yet worked out details of trade agreements with their neighbors. Where goods once flowed freely, now they halted at national borders. Everything was disrupted.

Inflation, crime, and shortages of goods increased. Unemployment rose in the cities. Ethnic wars broke out in Armenia, Azerbaijan, Chechnya, and elsewhere. Old scores were being settled, and thousands, eventually millions of refugees were on the move—with disastrous consequences in the countryside. Both Russian and North American specialists reported these consequences in books and scholarly journals, but few, if any, journalists followed suit. Stephen Wegren, for example, described a startling change in Russia's demographic profile, which until the end of 1990 had followed the trend noted in most industrial societies over the past 30 years, of net migration from the country to the city:

> New data on rural migration depict a reversal of rural outmigration. In complete contrast to the historical migratory patterns that predominated during the past 40 years, recent data on rural migration show that the countryside was a net recipient of migrants.[37]

He added that "the bulk of new arrivals during 1992–1993 were from other republics, presumably the near abroad [the term in the Russian Republic for neighboring states of the CIS] . . . my guess would be that a substantial number migrated from urban areas."[38] What might have influenced such an urban-rural movement? Dunlop, Ryvkina and Turovskiy, and Wegren—all writing in the scholarly, rather than the mainstream press—provided indications.

Dunlop was among the first to point out the "score-settling" nature of some events, noting that "there appears to be general agreement among specialists that a significant in-migration [of European Russian-speakers, largely from Central Asia] to Russia will take place over the remainder of the current decade . . . the pace of the migration process into Russia already has picked up considerably, beginning in late 1992."[39] He noted surveys taken among returnees, who cited "threats of violence and persecution" and "humiliation and abasement of their sense of national dignity" in the new republics, no longer politically dominated by Moscow, as reasons for returning to Russia.[40] The potential for in-migration was estimated from 800,000 to as high as 4 to 6 million persons.

"The existing practice," Dunlop added:

> has been to send the forced resettlers—most of whom come from urban areas, and often capital cities, located in

the "near abroad"—out into the Russian countryside. This bizarre attempt to turn engineers into tractor drivers and schoolteachers into milkmaids has not been a success. It also has led to the emergence of serious animosity on the part of the local rural population, who often view the newcomers as threatening their livelihood.[41]

He noted that when returning migrants sought permission to construct housing on the outskirts of large cities, they were turned down and shunted to "land located some 150 km away from oblast centres such as Kaluga and Voronezh."[42]

Ryvkina and Turovskiy gave additional details, explaining that: "Many of the refugees are skilled workers and highly educated professionals and intellectuals from large cities. Yet they are often forced to settle in rural areas where they remain unemployed."[43] The policy of resettling urban people in rural regions appears to be the result of a wrong-headed conviction by government officials that refugees should be shunted to collective farms, placed in special settlements they would construct for themselves, and that "they be permitted to settle in ecologically dangerous areas, such as Briansk and Orlovsk oblasti. . . . Most of the experts believed that refugees are most expediently settled in rural areas and in small, uninhabited towns or in district centres, where housing can be quickly and cheaply repaired and, most importantly, where it is easier to create jobs." The authors note, however, that "people do not in general want to go to such places."[44]

Wegren confirmed that "an overwhelming percentage of new arrivals" from the near-abroad "were Russian," but added that "a second main source of new rural arrivals has been former urban residents who have relocated to the countryside, that is, intra-oblast migration from urban to rural areas."[45] He added that "males with more than a secondary education were the most likely to both arrive and leave the countryside. . . . This trend suggests that those with relatively high education were among the first to depart to the countryside as urban conditions began to deteriorate."[46]

And he cited a significant increase in migration to rural areas of people aged 60 and above, adding that the trend "supports other information regarding how rapidly standards of living had fallen for pensioners since economic reforms were begun. It suggests that old people were moving to the countryside because they could not survive in cities with high food prices and extremely low pensions."[47] In short, things had become so bad in cities that people—both the educated young and the elderly—were forced back into the countryside.

It was unlikely, however, that the newcomers wanted to become full-time, commercial farmers. "There were no statistically significant correlations between rural immigration and the creation of private peasant farms, and it appears doubtful whether new rural migrants became private farmers. Furthermore, it appears extremely unlikely that people moving to the countryside joined agricultural enterprises (state and collective farms and their successors)."[48]

Migrants might plant subsistence gardens, but not fields of wheat. This was not surprising, given the fact that since post-communist agricultural reforms were introduced even rural residents who were working on state or collective farms have not flocked to private farming, preferring simply to expand their small private subsistence plots—most little larger than kitchen gardens.[49] As former Polish Deputy Minister of Agriculture Anna Potok put it, speaking of her own country:

> They [collective and state farm workers] got used to central planning—being told what to do—and have come to enjoy the benefits of social welfare. Now they are discovering that in a free market the means of production are very expensive. They say, "you've thrown us into deep water, but you haven't taught us how to swim!"[50]

The story was the same in Russia, as Shane noted:

> Despite the atypical encouragement from Ninkonenko, the state farm director, [private farmer] Gusenkov's former co-workers on the state farm next door had not rushed to follow his example and lease their own farms. Most of them seemed to prefer their shift work, their modest pay packets, and their occasional vodka binges to the responsibility of a private farm.[51]

A general pattern of populations on the move similar to that of post-Soviet Russia was noted in Depression-era Oklahoma by McMillan, who did his field work in the early 1930s:

> Generally speaking, in depression periods farmward migration increases because of the extreme difficulty of making a living elsewhere, while people already on farms tend to stay there for the same reason. . . . Large numbers of displaced oil-field workers and miners have moved back to the land. . . . Lacking the opportunity for employment in other extractive or manufacturing industries, miners settled on farms. There are still hundreds of unemployed miners waiting for the mines to reopen. . . . A less pronounced trend is the back-to-the-land movement induced by unemployment in urban centers . . . In [several] counties increases in farm population probably were caused by partially unemployed laborers being forced onto the land as part-time farmers in order to secure food, fuel, and housing at reduced money costs.[52]

Much like their Russian counterparts, many of Oklahoma's reluctant farmers had no real desire to farm full-time. As McMillan explained, they were "handicapped by a lack of capital and shortage of human resources. They have very inadequate teams, if any; they have neither cash nor credit."[53] In short, conditions in Russia after the fall of communism were much like those in North America during the Great Depression, with the CIS's own version of the "Okies" moving "down the road" in caravans as pathetic as those of Steinbeck's *The Grapes of Wrath*. In fact, the situation in the former Soviet empire was worse, complicated as it was by wars, ethnic feuding, and the collapse of the infrastructure (especially roads and railroads), rules, regulations, and enforcement mechanisms that had once held the unwieldy U.S.S.R. together.

Even now, after several years of IMF-imposed economic "structural adjustment," and some time for political ground-settling, the farm picture in Russia and much of the CIS remains bleak. Reports on the Russian Agricultural Listserv (RUSAG-L), an Internet discussion group for agricultural specialists, read like a litany of disaster:

- Russia's GDP fell three percent between January and May of this year, compared with the same period in 1995. Industrial production fell four percent, while light industry and construction materials decreased by a steep 24 percent.

120

- Ivan Gridasov, chairman of the crops department at the Ministry of Agriculture and Food, warned that this year's harvest may fall short of official expectations. Gridasov said the 1996 harvest will be difficult because of the drawn-out spring sowing and a sharp decline in harvest equipment. He explained that only 64 percent of machine demand is being met, which means two to three times more work than normal for every harvester, effectively slowing down harvesting and increasing harvest losses. Spring sowing covered [only] 58.1 million hectares of a planned 69.3 million.

- Government analysts expect Russian meat output to drop eight percent, to 5.5 million tons, and milk production to decrease by three percent, to 38 million tons of milk. In 1995, total cattle numbers were at 39.7 million, 17.3 million less than in 1990. Milk yields during the same period dropped from 2,781 kilos to 2,007 kilos. Russia needs approximately 4.8 million to 5.1 million tons of beef, but produces only 1.7 million to two million tons.

- The Moscow city government consumer market and services department reported . . . that Moscow's stores [including food markets and supermarkets] are still in bad shape because of high leases, taxes and community charges. Only the capital's largest and best known stores have a profitability of around four percent, while 650 smaller stores are on the verge of bankruptcy.[54]

Earlier, RUSAG-L had carried a still more ominous notice:

An economics ministry spokesman told ITAR-TASS that the Russian agricultural sector finished 1995 with a loss of two trillion rubles. Experts called rising prices of resources and services, declining buying ability of consumers, lack of competitiveness of many goods, and poor adaptation of farms to the market economy for the losses. The economics ministry official said that the pace of agrarian reforms has failed to alter production relations. He also voiced considerable concern over the growth of food imports, which reached 40 percent in 1995. Adding to that concern, experts at the Russian Institute for Studies of Market Conditions noted that

121

countries importing more than 30 percent of their food become dependent on swings in the world food market and imports.[55]

Around the non-Russian CIS, things weren't much better:

- Officials at Bulgaria's Agricultural Academy are predicting that the 1996 wheat harvest will not exceed 2.2 tons per hectare, down from 4.5 tons in 1991. This would result in the lowest yield in 20 years and aggravate an already acute grain shortage. Because of high world market prices, the government allowed grain exports last year, despite Bulgaria's own low grain supplies. The government was eventually forced to release emergency supplies and import grain from Romania, Serbia and India. Two agricultural ministers have already resigned this year over the grain crisis.
- According to USAID officials, the United States will give Armenia a total of 62,000 tons of grain this year under a $20 million aid initiative. Armenian officials said this latest agreement brings the total amount of U.S. aid to Armenia in 1996 to $150 million.
- Ukranian President Leonid Kuchma fired Agriculture Minister Pavel Haidutsky for allegedly failing to stabilize the financial crisis in the state-controlled agriculture sector, including failure to pay back-wages of 38 trillion karbovantsi (US$200 million). According to Prime Minister Pavlo Lazarenko, Haidutsky also failed to implement 12 policy orders, one of which provided for total price liberalization.[56]

A "Wild East" Atmosphere

Meanwhile, the post-Soviet press and broadcast industry have taken on an aura more reminiscent of the "Wild West" of American pioneer days than that of responsible, democratic institutions. In the "Wild East," journalists' troubles run the gamut from low pay, merciless competition and government censorship, to bombings, jailings, threats from Russia's out-of-control *mafiya*, and the brutal assassinations of both correspondents and press and television owners.

Although the attacks are not directed specifically at agricultural journalists, they have a strong chilling effect on news coverage in general. They force journalists to concentrate on immediate, violent political battles—as well as on their own personal survival—at the expense of thoughtful background coverage of specialist beats, like farming.

A sample of Internet postings from the East European Media and Cultural Studies List (EEMedia), the Former Soviet Union Media List (FSUMedia), and others shows that virtually every CIS republic is involved:

> LITHUANIA—A powerful bomb severely damaged a new wing of *Lietuvos Rytas*, Lithuania's largest newspaper, last week. The newly-formed Free Speech Foundation, representing a wide spectrum of the Lithuanian media, charged that the government's hostile attitude toward the press had encouraged the terrorist attack.[57]

> LATVIA—On the night of 27 September 1995, Inese Skutane, a reporter for the Baltic News Service, was abducted and later tortured. According to Skutane, three men attacked her, when she arrived at the location where she was to meet a contact who claimed to have material about Latvian Prime Minister Maris Gailis.[58]

> BELARUS—The Belarussian government forced three independent newspapers to suspend publication this week for what editors said was the threat they posed to the administration in next month's parliamentary elections.[59]

> RUSSIA—Russkaya Radio's broadcasts were interrupted at noon on 21 November after 20 men armed with submachine guns, some dressed in police uniforms, broke into the studio and damaged the transmitter, Russian media reported, citing chief producer Alexsandr Bunin. Bunin told Public Russian TV that the incident was the result of the radio's decision to refuse air time to an unnamed extremist right-wing politician.[60]

> ARMENIA—Three as yet unidentified young males, between ages 23 and 24, attacked Gagik Mkrtchian, a journalist in the employ of the independent *Golos Armenii* daily, on Wednesday morning (Sept. 27,

1995) as he was leaving his home. Mkrtchian is a former employee of the *Yerkir* daily newspaper, which was forcibly shut down by order of President Levon Ter-Petrosyan as part of his regime's large-scale campaign to silence Armenia's democratic opposition and independent media.[61]

BULGARIA—Bulgarian National Radio Director-General Vecheslav Tunev on 29 November dismissed two journalists working on BNR's "Hristo Botev" program, *Kontinent* reported the following day. Georgi Vasilski and Peter Kolev on 28 November broadcast a statement supporting journalists from Radio Horizont (BNR's other channel) who had recently issued a declaration accusing BNR's management of censorship.[62]

KAZAKHSTAN—All [independent television stations] report a certain amount of political intimidation, but the level varies greatly. The Shymkent station feels fairly free to cover whatever they want and to mildly criticize the city and oblast operations; but in the Ust-Kamenogorsk one station has had so many run-ins with the local government that its owner has a 24-hour armed bodyguard and has had shots fired into her apartment.[63]

MINSK—Belarus police were searching for a missing Russian television cameraman on Monday amid allegations from colleagues that his disappearance was politically motivated, the interior ministry said.

Dmitry Zavadsky, a cameraman for Russia's ORT television, was reported missing after his car was found empty at a Minsk airport on Friday evening. Other journalists said, "he had gone to the airport to pick up a colleague."[64]

The retirement of Boris Yeltsin from politics and the advent of new Russian President Vladimir Putin—a former KGB man—left little hope for improvement as the new millennium dawned. Putin, in fact, seemed determined to roll back the few advances in press freedom that had been made since the end of the U.S.S.R.

A pattern of intimidation and pressure on the Russian media casts doubt on President Vladimir Putin's stated intention to protect press freedoms, an international delegation of free press advocates said Thursday.

The group, speaking after three days of talks with government officials, journalists and publishers, said Putin has done little to back up his remarks that a free press is crucial for a healthy democracy.

"Putin's stated goals are far from reality in Russia today," said a statement from the Russian Press Freedom Support Group, made up of representatives from six international free press groups. "There is not a truly free and independent media in Russia."[65]

Only weeks before the report was released, Vladimir Gusinsky, the head of Media-Most, Russia's largest independent media empire, was arrested on apparently trumped-up fraud charges and thrown into Moskow's infamous Butyrskaya detention facility.[66] Though he was subsequently released, his arrest was widely interpreted as a warning from the new Putin government that it would not tolerate strong press criticism. Gusinsky's company controls Russia's leading commercial television station, NTV, which had been "one of the few vocal critics of Mr. Putin and the Chechen war."[67]

Along with violence and political repression, news media in the CIS face serious physical, economic, and social impediments. As in the West in recent years, newsprint costs are a chronic problem, made worse by Russia's current disorganization. Nicholas Pilugin, of the USIA Media Assistance Clearinghouse in Moscow, reports:

> While the price of Russian newsprint has been rising and it has hurt newspapers, just obtaining newsprint is the greater problem. Another issue is the high tariff charged on newsprint imported into Russia.
>
> Newsprint has recently cost about $700 a ton. But the same Russian newsprint can be purchased for delivery in Paris at $450 a ton (this according to the Russian Union of Journalists). But the customs duties are such that if you try reimporting it, it will cost you even more than $700 a ton.

I recently visited Vladivostok, where I met with the editor of one newspaper there. He told me that in the Far East there are two newsprint plants, only one of which approaches anything near reliable delivery. But deliveries from even that plant are spotty, and the editor indicated he had had to cut back the size of his newspaper to conserve his stocks—while a newspaper in Khabarovsk was in danger of temporarily shutting down due to a lack of paper.[68]

These and other economic problems have come close to crippling the print media in the Russian Republic, as the Open Media Research Institute reveals:

According to a seminar on democracy and the Russian mass media held in Moscow, circulation of newspapers and periodicals in Russia is only at seven percent of 1990 levels, Russian TV reported on 5 July.[69]

Further:

More than 85 percent of Russian publications are not financially independent, according to Iosif Delashinskii, head of the analysis department at the State Press Committee, Radio Rossii reported 10 July. There are now 10,500 newspapers in the country, most with a print run of less than 10,000 copies. Delashinskii said that as much as 70 percent of the country's printing equipment needs to be replaced.[70]

As for Russian television, it has its own problems, some of them quite basic. For example, take the concept of "pay-TV," so common in the West. Asked about it, Ukrainian Mykola Kniazhytsky, director of the International Media Center/Internews, in Kiev, replied simply: "As far as our people are concerned, the majority are not able to pay for food, so it is understandable that they won't be able to pay for TV."[71] The International Labor Organization (ILO) estimates that more than 100 million people in the former U.S.S.R. live below the poverty line. In Moscow in 1992, observers estimated that the "average household was spending over

75 percent of its income on food alone—and macroeconomic conditions have generally worsened since then."[72] In such a difficult environment, providing regular, in-depth coverage of any subject area at all is a major achievement, and it is not surprising that hard-pressed journalists' immediate attention should be focused on the high-profile political crises of the moment. As Mikola Kniazhytsky noted in reference to television news coverage in the Ukraine:

> The information that our audience receives about events from around the country is limited to what comes out of metropolis [Kiev], or I would even say limited to what is happening in the three government establishments: the president's administration, the Supreme Rada and the Cabinet of Ministers. All news and analytical programs highlight events from these three establishments. Of course, it is important and interesting, but it is not enough.[73]

Agriculture tends to get lost in the shuffle, as Slovakia's Viera Simkova, quoted in the Introduction, explains:

> In Slovakia [population 5.3 million], a small group of active reporters is covering agriculture. There is also a special daily for this issue called *Rolnicke Noviny* (*Agriculture Daily*). . . . There are not a lot of articles in the media about agriculture at this time, because it seems old-fashioned to write or speak about these items. It was very common in the former communist regime, when newspapers, radio and TV were over-full with it, so one can observe the opposite now.[74]

To which Nicholas Pilugin, referring to Russia, adds: "There used to be a special agricultural department in TASS [news agency] in the Soviet period. Now [at its successor agency ITAR-TASS] there are [only] two to three reporters."[75] He notes that, as in Slovakia, there is also a newspaper dealing only with agricultural issues, *Krestyanskiye Vedomosti*, but it is directed at specialists.

Hilmi Toros, chief of the press section of the United Nations Food and Agriculture Organization (FAO), has also noticed a change:

> We used to reach the U.S.S.R. and East Bloc via the news agencies, such as TASS and TANJUG. We would send material to them and they would pass it on to the specialized agriculture magazines and so forth. We got fairly good coverage. We still send to them—TASS is now ITAR-TASS—but we don't get any feedback. We don't know if there are any farm magazines, or who they are. They may not have started up yet.
>
> When you look at the number of farmers in the area, there is certainly a potential growth area for private companies to start farm publications.[76]

A few Western groups have attempted to fill the vacuum left by the shrinkage of the state's media apparatus and the failure of struggling local independent media to find their feet. For example, the U.S.-based Rodale Institute, publishers of *Organic Gardening* and *New Farm* magazines, has launched a Russian-language counterpart of *New Farm, Novii Fermer.* The institute, founded by the late organic farming exponent Robert Rodale and continued by his son, J.I. Rodale, is one of the world's leading research facilities for organic horticulture and sustainable agriculture. However, the efforts of *Novii Fermer* are not focused on informing the general public about the issues in agriculture, but "on getting readers started in what Americans would call market gardening. 'The magazine for everyone who works their own land' is how it's described on the cover."[77]

The same is true of the farm radio initiative of Associate Professor Janet Macy, of the University of Minnesota Department of Extension Communication. Macy set up several focus groups in the former Soviet republics and broadcast Russian language farm radio programs from a studio in Nadezdah. The object of the project was "educating farmers to operate private instead of public farms."[78]

Educating the general, voting public about agricultural issues that affect their quality of life does not appear to be the object of any of the publications or broadcast programs supported by aid initiatives currently in place in the former East Bloc.

Journalism Education

The Association for Education in Journalism and Mass Communication (AEJMC) annual directory lists officially sanctioned journalism/mass communication schools in the following East European and CIS states: Albania, 1; Belarus, 3; Bulgaria, 2; Croatia, 1; Czech Republic, 1; Estonia, 1; Hungary, 2; Latvia, 1; Lithuania, 1; Moldova, 1; Poland, 6; Romania, 3; Russian Republic, 5; Slovakia, 1; Slovenia, 1; Ukraine, 4.[79] Queries directed to these schools failed to locate a single program in agricultural journalism. In fact, letters directed to the five Russian schools elicited no response at all (perhaps an indication of the poor state of the new republic's postal services). The American University in Bulgaria, a joint Bulgarian-American institution that has a regular program in journalism and mass communication, indicated interest in "working with media professionals in rural areas."[80] However, it did not have any courses in place oriented toward farm journalism.

In addition to indigenous journalism faculties, a number of western institutions are or have been engaged in journalism education projects in the former East Bloc. For example, the U.S. government sponsored the International Media Fund, a nonprofit, five-year project to aid in development of independent media in Central and Eastern Europe. Plans in late 1995 called for it to be succeeded by a 36-month joint U.S. Agency for International Development (USIAD)/non-government organization (NGO)-supported Professional Media Program, emphasizing "a heavy business management approach" aimed at helping independent media to become financially self-sustaining. One of six projected activities under the program was to have been "university journalism education." For the program's first year, USAID was to have spent $1.6 million in a startup program including Albania, Croatia, Hungary, Lithuania, Slovakia, and Romania. Poland, Bulgaria, Bosnia, and Macedonia were to be added later. However, spokespeople for USAID warned that the program was being announced "against a background of budgetary uncertainty," as there is "a strong movement in the U.S. Congress to drastically reduce foreign aid and transfer USAID to the State Department."[81] USAID noted that the program, if funded, would emphasize "outreach to media in the provinces,"[82] but listed no specific courses in farm journalism. A $10 million Russian-American Partnership Program (RAMP), funded under separate legislation, was to conduct "roughly equivalent" training in Russia.

The European Journalism Center in Maastricht, The Netherlands, also planned to "look into possibilities to propose project support for training rural journalists." A mid-career training center and clearinghouse for media projects of European journalism schools, the EJC noted that it "plays a key role in (re)training CEE/CIS journalists" and is "involved in the restructuring of many Eastern European J-schools. [83]

In addition to such programs, there are a number of smaller projects dealing with journalism in the former East Bloc. The U.S. Information Agency (USIA), for example, planned to offer four-week fellowships to television professionals from Russia and the Ukraine on "the role of television in a democracy."[84] Co-sponsored by USIA, Duke University, and the International Commission on Radio and Television Policy, the fellowships were to be offered in 1996 and 1997. Professor Jay Brodell of Metropolitan State College of Denver has even developed a Russian-language instruction booklet for classified advertising salesmen.[85]

However, such projects contain no specifically agricultural component. Nor is their continued financing assured in a period of "structural adjustment" and increasing disillusion in the West with the not-so-new Russia and its former satellites. Journalists in many ex-Soviet states must continue to plead for western support for education and training:

> Seventeen media NGOs from Central and Eastern European countries recently urged national and international donors to continue their support for media training in the region in order to further strengthen the democratization process, the European Journalism Center (EJC) reported.
>
> Represented at a late June roundtable sponsored by the Dutch government at the EJC in Maastricht, Netherlands, were NGOs from Albania, Belarus, Bosnia, Bulgaria, Croatia, Czech Republic, Estonia, Hungary, Latvia, Lithuania, Moldova, Poland, Romania, Slovakia and Ukraine.
>
> Participants stressed that more training of journalists, editors and media managers is imperative.[86]

While this survey is not exhaustive, it should give an indication of the scarcity of training available for agricultural journalists in the region.

CHAPTER EIGHT

Why Not Africa?
Farm Journalism in the South

*In mass communication, as in most other
aspects of modernity, Africa is slipping further
and further behind the rest of the world.[1]*
—William A. Hachten

Every news broadcast on Voice of Kenya (VOK) Radio started the
same way: A recorded brass-band rendition of the Kenya national
anthem, followed by the inevitable:

"Good evening (pause). His Excellency the President,
Daniel T. Arap Moi, lunched today with members of the Embu district KANU executive . . . " or "Good morning (pause). His
Excellency the President, Daniel T. Arap Moi will lunch today with
members of the Naivasha city KANU executive. . . . " or "Good
afternoon (pause). His Excellency the President Daniel T. Arap Moi
today warned members of the KANU party to be ever-vigilant
against corruption . . . " For three-quarters of each broadcast, the
most banal activities and/or pronouncements of Kenya's dictator—
most often something connected with Moi's Kenya African National
Union (KANU) party—would be recounted in minute detail. Then,
as a sort of afterthought to fill in the last few seconds before the
news break ended, one or two other items—the landing of U.S.
marines in Panama, say, or the destruction of the Berlin Wall, or perhaps the discovery of a cure for a major disease—would be thrown

in, in no particular order. Then the national anthem would sound again, and Kenyans would be presumed to have heard all of the information their government-owned radio thought they needed.[2]

The broadcasts were the butt of endless jokes among expatriates working in the country, who mocked the news readers' delivery: "Good evening (pause). His Excellency the President, Daniel (sound effect) Arap Moi today drank too much *pombe* (beer) before lunching with (sound effect) members of the KANU executive. . . . " Everyone had a good time with it, and no one took the broadcasts, or VOK, very seriously. No one, that is, except those who reflected on the fact that a valuable communications resource, which could have been employed to much more useful effect in a country beset with daunting social and economic problems, was being wasted in such crude, ineffective propaganda. It is one of the tragedies of Africa, at the close of the twentieth century, that VOK is not unusual, but typical of the quality level of official news media throughout much of the sub-Saharan region. Independent media sometimes try to rise above this level, but their existence is in constant peril.

To understand why this should be so, African journalism, particularly farm journalism, must be seen in context.

Altered Alignments

With the collapse of the Soviet system, the rise of Asia's economic "tigers" (Singapore, Hong Kong, Taiwan, and South Korea), and the coming of new trade pacts and a measure of democratic development to several countries of Latin America, it is no longer possible either politically or geographically to use the term "Third World" as a blanket descriptive. Too many alignments have been altered; the "Second World" or East Bloc doesn't exist anymore, and many countries once deemed "underdeveloped" are now among the globe's rising economic powerhouses. Asia's original tigers, for example, have been joined by Thailand and Malaysia, with Indonesia pressing at their heels. Costa Rica, once listed as a developing country, has progressed to the point where bilateral aid infusions are almost unnecessary. A decade ago, the U.S. foreign aid office in San Jose "was handing out $285 million; now the figure is $15 million, and the next benefactor to go will probably be the Peace Corps."[3]

Yet if any region of the Global South still wears the face of poverty so uniformly as to fulfill the definition of the "Third World" as a place of deprivation, it is Africa south of the Sahara. In the 44 sub-Saharan countries, most of the stereotypes associated with underdevelopment still hold true. In many cases, they are worsening. Of the world's 34 least developed countries (LDCs), 26 are located here.[4] The region's population of some 131.6 million people represents roughly 11 percent of the world population, and more than half are under 15 years of age.[5] According to the FAO, economic expansion in the area "in 1994 still remained well below population growth, bringing the region's cumulative decline in per capita income to 15 percent over the past 20 years."[6] This economic decline, Merrill adds:

> has been further exacerbated by the lack of adequate health care systems, low literacy, high unemployment rates, and alarmingly high teenage pregnancies. According to a 1993 report by the Fellowship of Christian Communicators in South Africa, AIDS is systematically wiping out a whole generation of young people in Africa. It is estimated that, by the end of this century, there will be more than three million AIDS orphans in sub-Saharan Africa. Compounding this problem is the emergence of drug-trafficking, money-laundering and the proliferation of all types of weapons that have destabilizing and debilitating consequences in African states.
>
> One of the results of this situation is that private investment flow to African countries has dropped from US$10 billion to $4 billion in the early 1990s. Moreover, new loan commitments to this part of the world by the World Bank in 1993 fell by $1.2 billion (or 30 percent) as compared to an increase from $1.7 billion to $3.8 billion for Eastern Europe during the same period.[7]

As for agriculture—of primary interest in a discussion of agricultural journalism—the FAO reports that:

> In the agricultural sector, which typically contributes one-third of GDP and employs more than two-thirds of the economically active population, the last 25 years have been characterized by steadily declining per capita pro-

duction. From a peak in 1975, per capita agricultural production had by 1993 declined by 20 percent, with a brief interruption of the downward tendency only during the second half of the 1980s. The pattern for food production has been almost identical, but with the decline in per capita terms over the same period reaching 23 percent. After a modest recovery in 1993, the downward trend in per capita production volumes was resumed in 1994.[8]

At the same time that per capita production has been declining, the value of agricultural exports from the sub-Saharan region has also been dropping steadily. A graph showing the overall quantity of agricultural exports against their value and unit value from 1987 to 1993 shows a V-shaped divergence, the value declining in roughly inverse proportion to the increase in quantity.[9] In short, the more agricultural produce Africa sells, the less money it gets. As for future prospects, the FAO notes that "gradually declining prices for Africa's main agricultural export products are expected."[10]

The industry that employs two-thirds of the continent's people is in a tailspin, spiraling downward. Why should Africa be faced with such bleak economic and social prospects?

As already outlined in chapter 3, part of the blame can be traced to the impact of the global debt crisis of the 1980s and its depression of farm prices, to the draconian terms of the World Bank/IMF-imposed structural adjustment programs that followed it, and to the expected effects on North-South trade of the more recently concluded GATT agreement. Also still playing a significant part are the long-term effects of colonialism—not least among them the European occupiers' creation of artificial and unrealistic national borders that ignored the established distribution of Africa's cultural and language groups and paved the way for today's viciously destructive intertribal power struggles and ethnic wars.

To these external, or externally-imposed circumstances are added a home-grown handicap, as Merrill explains by quoting Gurirab:

> After some 30 years of independence, the peoples of Africa . . . "have been neglected and impoverished by

their own governments amidst an abundance of natural resources and wealth being siphoned off by their—in most cases—unelected leaders." (Gurirab 1994, p. 5).[11]

From Idi Amin and "Emperor" Bokassa in the 1970s to Daniel Arap Moi and Robert Mugabe today, Africa's dictators and their corrupt entourages have pillaged their countries' economies and stifled political and social development on every front. Agriculture and rural people have frequently suffered the noxious effects.

In Kenya, Kikuyu farmers in the Great Rift Valley were deliberately targeted by roving gangs of thugs—widely believed to be out-of-uniform soldiers—and terrorized during the runup to elections in the early 1990s.[12] The Kikuyu were seen as anti-Arap Moi voters, and despite their reputation for being among the best farmers in Kenya, were not wanted in the district. Many were murdered and others were run off, and their homesteads burnt.[13]

In Robert Mugabe's Zimbabwe, white farmers became the target of vicious electoral politics during the spring 2000 elections, in which Mugabe tried to curry favor with black rural voters by instigating and subsequently approving squatter takeovers of the most successful white-owned farms. No compensation was offered the expropriated whites, most of whom were native-born Zimbabweans, and the squatter occupations were frequently violent. At least five farmers were murdered and many others were threatened:

> Squatters accosted scores of white farmers in Zimbabwe Friday and ordered them to leave their properties as President Robert Mugabe fixed today as "D-day," marking the "final phase" of his land grab.
>
> Under pressure to reward the squatters, who became the shock troops of his Zanu-PF party by mounting a brutal onslaught against the opposition during last month's election campaign which claimed 37 lives, Mugabe has listed 804 farms for "compulsory acquisition."
>
> About 1,200 farms are currently occupied, and the increasingly impatient invaders are hammering at farmers' gates across Zimbabwe and ordering white families to leave.

At Woodlands Farm, near Shamva, 112 km north-
east of Harare, 32 invaders were massed Friday at the
gates. Their leader, incoherently drunk and wearing a
large knife at his waist, had given Keith Butler and his
brother Mark, three days to leave.[14]

Crippled Media

Many of the same handicaps that have affected African development
generally have also played key roles in crippling the growth of sub-
Saharan news media. The legacy of colonialism, for example, is sum-
marized by Merrill:

> The mass media in most African countries during the
> colonial era were used by settlers to promote their ideals.
> The media mostly ignored the local [African] population,
> and focused on events happening in the settler commu-
> nity (Hachten 1993, p. 17; Martin 1991, p. 159).
>
> Europeans controlled the economy of colonial Africa,
> and by implication they had monopolized the media dur-
> ing this era because "few Africans could afford to import
> the necessary equipment for printing newspapers or set-
> ting up radio stations" (Ziegler and Asante 1992, p. 26).
> It was also deliberate colonial policy to allow Africans lit-
> tle access to the capital needed to get newspapers or estab-
> lish radio stations. The colonials denied the African press
> rights because they believed that the press in the hands of
> Africans was a potentially powerful tool against colonial
> rule. Thus, colonial administration in Africa could be said
> to have bequeathed contemporary African governments
> with draconian press laws and a legacy of government
> monopoly of the media. Some of these colonial press laws,
> unfortunately, still exist in some African countries. In
> some instances, these press laws, designed to curtail press
> freedom, have been put in practice by authoritarian
> African governments and have assumed even more sinister
> dimensions.[15]

Colonial policies were particularly damaging in West Africa,
where:

The French colonial policy of assimilation hindered newspaper growth and development in the former French West African colonies. Also hindering media development were low literacy levels arising from the French policy of selective education, the French requirement that prospective newspaper publishers be French citizens who were in good standing, harsh taxes on importation of newsprint into the colonies, strict press laws, and Paris as the hub of intellectual activity. The press, as a result, remained an appendage of France and did not actively seek or encourage African readers.[16]

The poorly performing economies of Africa create another brake on media growth by impoverishing the commercial advertising base on which private publications and broadcast stations must depend for survival. As Peter Mwaura of the United Nations Educational, Scientific and Cultural Organization (UNESCO) explains, discussing the media in Kenya:

> The economics of publishing dictates that no commercial newspaper or magazine can stay alive unless it generates enough advertising revenue to pay its way. A few newspapers, magazines and journals that carry little or no advertising exist in many parts of the world, but they are often sustained by a high cover price (which makes them out of reach of ordinary newspaper readers) or enjoy some form of subvention. The growth of newspapers and magazines in Kenya has been dictated by such economics of publishing. Normally a country can have only the number and size of newspapers and magazines that its economy—or more specifically its retail business and the willingness to advertise—can support.[17]

In the years following independence (1963) under Jomo Kenyatta and the initial period of Arap Moi's reign, Kenya's economy was growing, external development aid was abundant, and the conditions for publishing were relatively good compared to the rest of the continent. As Mwaura writes:

> In the developed nations, such as the United States of America, a newspaper normally has to carry a 65 percent

and 35 percent advertising-news ratio to make a profit to enable it to stay alive. But in many developing countries such as Kenya where operating costs, such as wages, are generally lower, the advertising-news ratio differs. In Kenya this ratio varies between 30 and 35 percent advertising. It is also generally believed that, compared to developed countries, only an infinitesimal fraction of Kenya's gross national product is used for advertising and promotion; the country could probably support far more newspapers and other media than it does at the moment.[18]

However, the increasingly oppressive corruption and despotism of the Moi regime; deteriorating world prices for agricultural products, including Kenya's two leading exports, coffee and tea; and a general cutoff of external aid by the world's leading donor nations (prompted as much by the fall of the Soviet Empire, and the end of the need for client states, as by Moi's corruption) have pushed the country's economy into decline. It is doubtful that Mwaura's optimistic assessment of advertising prospects could be made in Kenya today. In other sub-Saharan countries, which have never enjoyed Kenya's level of relative prosperity, the advertising base for independent print media is even weaker, as Hachten notes:

The economic and social conditions prevailing, for example, in such capitals as Bamako in Mali, Addis Ababa in Ethiopia, Conakry in Guinea, Niamey in Niger, Freetown in Sierra Leone, and Kampala in Uganda, during the past 25 years have not been hospitable to fledgling newspapers of any sort. Neither African one-party nor military rule permitted free and unfettered political discourse, nor did the economies offer much promise of return on investment.[19]

As for broadcast media, most radio and television systems in Africa are government owned or dominated by a ruling party. Observes Merrill:

Political and government leaders feel it is not wise to allow radio and television companies in the hands of those who can challenge state policies, or have alternative perceptions of the situation in the country."[20]

Not only is the independence of radio and television compromised but also their expansion is blocked by limited government budgets—becoming ever more limited under World Bank/IMF structural adjustment policies. As Merrill notes:

> Not much attention is paid to financing the media to enable them to operate smoothly. Journalists are among the most lowly paid people and use outdated and obsolete equipment. Transport to ferry journalists to assignments is nearly always in short supply.
>
> The root cause has been the token funds African governments allocate to official media in their annual budgets. In one country, the money allocated to run the broadcasting house for the whole year was not even enough to buy one small transmitter, leave alone maintain existing ones, which were in very bad shape. Little wonder this particular country had to go off the air for lengthy periods during the year because the spares for the shortwave transmitters could not be purchased. The problem with many African governments is that they rely too much on foreign aid money to run their information departments, particularly broadcasting. When donor money is not forthcoming, the state of the government-owned media remains in shambles.[21]

Government censorship even restricts potential advertising revenues. In Kenya, for instance, Merrill observes that:

> Since 1959, commercials in the Kenyan media have to be approved by the government on the basis of content, length and timing. Subsequently, the revenue the [broadcast] media get from commercials is rather small (Abuoga and Mutere 1988, p. 114; Heath 1988, p. 101)."[22]

To which Mwaura adds:

> Most of the advertising is in the English language and reaches a small group of well-to-do and educated people through newspapers, magazines, cinemas, radio and television. The [Ki]swahili press is the Cinderella of advertising in Kenya, a fact which has hampered its growth in

a country where [Ki]swahili is the national language and the lingua franca, and is, according to research, preferred by the majority of Kenyans.

Television, which is controlled by the Voice of Kenya and reaches a far greater number of people than the print media, is also ignored by the advertisers in spite of its potential impact as an audio-visual medium and captive audience. Perhaps the explanation for this is that most of the television audience can be reached through the print media, at least the audience that matters to advertisers; it is also possible that the potential advertisers hold a poor view of television programing in Kenya and are reluctant to support a medium they do not like.[23]

In addition to the above problems, there are a number of other barriers to media success in Africa. Hachten summarizes them:

Increasing illiteracy, inadequacies of education [in many cases, education budgets had to be sharply cut to meet the harsh dictates of WorldBank/IMF structural adjustment plans], widespread poverty, endemic disease (complicated by the devastating new plague of AIDS), and malnutrition—all of this means that except for a few urban elites, the potential readers and listeners for the media were just not there. If a new paper is launched, the problem is how to distribute it outside the city of publication when mail service is so erratic and the road and railway systems so chaotic and unusable.

Private capital to launch new journals and papers is largely nonexistent. Most of the best independent newspapers throughout the world have flourished in free and open societies where private property was protected by law and where newspapers were profit-making ventures in market economies. In those mostly Western (or Westernized) societies, investment capital was available to the publishing entrepreneur willing to take the risk of putting out a newspaper.

But in most African societies, with few exceptions, such economic conditions and entrepreneurs were notably lacking.[24]

In placing illiteracy and inadequate education at the top of his list of impediments, Hachten recognizes reality. Newspapers and magazines are of no use to people who can't read them, and in Africa illiteracy is a severe problem. While a few countries—Kenya, Madagascar, Zaire, Zimbabwe—have male literacy rates above 80 percent, 14 sub-Saharan countries have male literacy rates below 50 percent, and several others hover just above the halfway mark.[25] Even those with a majority of literate males have much lower female literacy rates. In Burundi, for example, while 61 percent of adult males can read, only 40 percent of adult women are literate.[26]

Even those who can read may find the task insurmountable if the periodical in question is in the wrong language. As Martin Ochs explains: "In only 29 of 51 African countries are Arabic, English or French a factor. The following are only a few of the continent's 800 languages and dialects: [he then lists the major languages of 26 countries, from Malagasy and Chichewa to Temne, Mende, Creole, Hausa, and Yoruba]."[27] He points out that:

> In black Africa's most highly developed media system, that of Nigeria [which boasts no fewer than 200 local languages], none of the daily newspapers is printed in other than English. In much of sub-Saharan Africa, [Ki]swahili is just about all the non-English press there is.[28]

Even a literate but unilingual Yoruba speaker will be left out if the only publications available are in English.

As for educational level, which determines the ability of potential readers to fully understand and appreciate what they read, sub-Saharan Africa falls behind every other world region. In Burkina Faso, only 25 percent of youngsters in the relevant age group are enrolled in primary school, only 4 percent in secondary school, and a negligible 0.6 percent in post-secondary classes. Compare this to India, where 98 percent of children are enrolled in primary school, 39 percent in secondary school and 6 percent in post-secondary school.[29] There are, of course, African exceptions. Mauritius has 92 percent of its youth in primary school, 47 percent in secondary school, and 2.1 percent in higher education—but it is an obvious anomaly.[30]

To which Ochs adds that:

In such a climate can be seen the difficulty of establishing the world's most varied literacy efforts, but many of these are succeeding. The positive side of a one-third African literacy rate is that this rate has at least doubled since 1976. The progress is the more remarkable since a high percentage of African languages have no written form. Projects to construct an alphabet in some of these are among the most interesting of all African development. Still, it will be a long process.[31]

Selling newspaper and magazine subscriptions to people who have no money is another impossible task. Even in a comparatively well-off African city like Nairobi, the average African resident rarely earns more than the equivalent of US$20 per month. Many are lucky to earn half that. The $35 to $45 price of a year's subscription to a European or North American consumer magazine is obviously out of the question for all but a wealthy elite. As for local publications, quality magazines published in Nairobi, such as *The Nairobi Law Monthly*, at 50 Kenya Shillings per issue (roughly $1.60 in 1992)[32] or the *Monthly News*, at 30Ks ($1) per issue are also beyond the average person's reach.[33]

The brisk sidewalk trade in used, often stolen, magazines and newspapers (every city block has three to five secondhand magazine vendors sitting on the pavement on each side of the street, or hawking their wares to drivers at the traffic lights) is graphic testimony to the true value of reading material in print-starved Africa. A single copy of the day's newspaper will be sold, resold, and resold again, passing from hand to hand for a week or more until it is in tatters. Yet only the first newsstand sale generates revenue for the publication itself. All subsequent exchanges are strictly informal.

As for private capital to launch new journals and papers, as Hachten has already been quoted as pointing out, such funds are "largely nonexistent." In countries like Kenya, it is mostly the "Asian" (Hindu East Indian) and Muslim minorities that dominate business and have sufficient funds to invest in new ventures (a majority share in both the English-language *Daily Nation* and Kiswahili *Taifa Leo* newspapers, for instance, is owned by Aga Khan IV, the Persian/Indian Imam [head] of the Ismaili sect of Shia Muslims, whose family invested in the papers years ago). Yet this group is among the most socially and politically insecure in the subcontinent. Explains Patrick Nagle:

The East Indian and Muslim communities [are] warned regularly by inflammatory KANU demagogues that they are in the country on sufferance and can be ejected or sequestered at any time. . . . The Asians are carefully aware of their uneasy relations with the black majority and miss no opportunity to subscribe to any KANU fund-raising event, although privately they say they are being blackmailed.[34]

No one in these communities has forgotten former Ugandan dictator Idi Amin's brutal expulsion of his country's Asian minority, and the pillage by his thugs of the property the Asians left behind. To expect such a vulnerable minority to support a politically critical, independent press is simply unrealistic. Describing Kenya's dailies, Hachten observes that some:

do a creditable job of reporting African as well as world news. What they do not do well is report on the domestic politics of Kenya: they are susceptible to political pressures and exercise much self-censorship. The papers do not report in depth or criticize President Daniel Arap Moi, just as they did not attack [Kenya's founder and first President Jomo] Kenyatta or any members of his family. In 1982, the editor of the *Standard*, George Githii, was fired because of an editorial that criticized the government for detaining people without trial. The editor of the *Daily Nation* suffered a similar fate on another issue. . . . The outlook for Kenya's relatively free press and its fragile democracy is not promising.[35]

Since Hachten's book appeared in 1993, political repression of the independent press in Africa has increased. A glance through some recent Internet postings for African discussion lists gives a picture of what—thanks largely to Africa's own leaders—is happening to the press.

LIBERIA—Liberian police stormed the *Inquirer* newspaper Thursday, but passers-by and staff hurled insults and prevented them from seizing production manager Jacob Doe after the paper published a front-page story criticizing the arrest of two editors of the *Daily Observer*.

Police Tuesday detained *Daily Observer* editor James Seitua. When managing editor Stanton Peabody went to seek the release of Seitua Wednesday, he also was arrested.

Authorities announced Thursday the two editors had been charged with criminal offenses for articles their newspaper carried last week alleging presence of mercenaries in the capital with links to rebel leader Charles Taylor.[36]

ZAMBIA—President Chiluba has rapidly distinguished himself as one of the worst violators of press freedom in southern Africa. Using criminal law as a form of censorship to protect himself against unfavorable criticism, President Chiluba has routinely charged the independent press, namely the leading Zambian independent daily *The Post*, with libel. Employees of *The Post* have been hauled into court for their reports on governmental corruption so often that they have lost count of the numerous charges filed against them, the sentences for which amount to more than 100 years in prison. This year, President Chiluba banned the February 5 edition of the newspaper as well as that day's on-line edition, marking the first act of censorship on the Internet in Africa.[37]

NIGERIA—Nigerian security agents raided the homes of *Tell* magazine editors Nosa Igiebor and Onome Osifo-Whiskey Sunday and confiscated the entire press run of the latest edition before it hit newsstands.

Agents of military ruler Gen. Sani Abacha confiscated the entire print run of *Tell* magazine's weekly edition, scheduled to hit newsstands Christmas Day with a cover story entitled "Abacha is adamant." Igiebor was picked up at his home and taken to State Security Service headquarters.

Osifo-Whiskey, in a statement released from hiding after his home was raided, said a dozen security agents went to the magazine office and demanded to see "senior journalists." They forced doors open and left after a search of the premises, he said.

Last week agents went to the same printing press and confiscated 55,000 copies of last week's edition. When the magazine sought a different printer, agents went through Lagos harassing vendors and grabbing copies off the newsstands.[38]

COTE D'IVOIRE—Three journalists, Abou Drahamane Sangar, Emmanuel Kor and Freedom Neruda of the independent media group Nouvel Horizon were sentenced to two years imprisonment in January for a satirical article in the daily newspaper *La Voie*, which stated that President Henri Konan's presence at a championship football game brought bad luck to the national team.

Since President Konan took office in December 1993, there has been an increase in prosecutions against journalists sympathetic to the opposition which have resulted in sentences of up to three years imprisonment. Attacks on independent Ivoirian journalists also come in the form of physical harassment: In June 1995, Mr. Sangar was savagely beaten by four policemen at the order of Security Minister General Ouassenan Kone for an article he wrote in his satirical weekly, *Bol Kotch*, about the minister's treatment of student unrest in the country. The beating took place in Minister Kone's office, and in his presence.[39]

ANGOLA—Threats against journalists, especially the threat of murder, are unwavering. Since the signing of the Lusaka Protocol in November 1994, it is evident that neither side in Angola's bloody civil war has observed the peace plan's terms, particularly press freedom. Barely two months after the signing, editor Ricardo de Mello was gunned down in front of his house. His assassination brought the number of killed journalists in Angola to 12 since the September 1992 elections.[40]

KENYA—Attacks on the independent media continue unabated, so much so that the Committee to

Protect Journalists (CPJ) has named President Moi one of the world's top 10 "Enemies of the Press." President Moi has declared war on the press and has widened his net to include foreign correspondents. Critical coverage of Moi's administration has been declared a criminal offense, and newspapers or printers have been arbitrarily closed for publishing opposing viewpoints. Journalists covering the trial of human rights activist Koigi wa Wamwere have been physically attacked.[41]

CAMEROON—In the very country where African heads of state are currently gathered, prior (pre-publication) censorship continues to be practiced—one of the last places on earth to do so. Last year CPJ recorded at least 10 incidents in which independent newspapers were seized or banned. Vendors have been robbed or beaten by police, citizens harassed for reading banned papers, journalists arbitrarily detained without charge, interrogated and assaulted. Radio and television broadcasting remain under complete control of President Biya's ruling party.[42]

BURUNDI—French journalist Jean Helene, questioned for two hours in Burundi about discrepancies between his professional name and passport then released, was denied permission to leave the country Thursday in an effort to discourage foreign news coverage of the country.

Helene, a reporter for the daily *Le Monde*, has received numerous death threats for articles critical of Burundi's ethnic Tutsi-dominated military. He has covered Burundi for more than five years. Helene attempted to fly back to Nairobi, but was detained at the airport by security forces who searched him and confiscated his passport. The French ambassador intervened on his behalf, but Helene was not permitted to leave.

"They know who I am and I know I am not popular here," Helene said. "They are trying to send a message to international journalists."[43]

It is not surprising that all of the preceding reports of censorship and violence involve print media, since, as already noted, most broadcast media in Africa are government-owned and toe the "party line" of incumbent officials.

How Africans themselves feel about this situation was expressed in the summer of 2000 by a member of an Internet discussion group on Africa. He was reacting to complaints about perceived "racism" in BBC reporting of African events:

> Look, to solve a problem you expose it, diagnose/analyse it, prescribe a solution, implement it and monitor it. To expose a problem, you need a MEDIA and to diagnose, analyse and debate solutions to a problem, you need a FORUM.
>
> Now, tell me this: in how many African countries can we EXPOSE problems of corruption, violations of human rights, repression, civil strife, etc.? In the STATE-OWNED MEDIA? And when a problem is exposed, in which FORUM is it analysed? In the RUBBER-STAMP parliaments?
>
> These are the areas where we should focus our attention—not the BBC. Because if we Africans had been exposing our problems and debating our own solutions for them, there would be no need for the BBC to ask racists to do it for us. Do you agree? Then leave the BBC alone.[44]

Beyond City Limits

All of the impediments facing the media in urban Africa are multiplied when a publication or broadcast station tries to reach beyond the city limits, or to cover the news in rural districts. Illiteracy, for instance, is generally higher in rural than in urban areas, and the already mentioned tendency of African female literacy to lag behind that of males has particularly adverse effects in the countryside. According to the FAO's Jennie Dey, women provide "60-80 percent of the labor in food production and a substantial contribution to cash [crop] production in many African countries."[45] If 50 to 60 percent of the women in most of sub-Saharan Africa are illiterate,[46] then the majority of Africa's farm workers are incapable of reading a newspaper or magazine article about agriculture.

Generating sales outside of major cities is even more difficult. Along with lower literacy rates and lower disposable cash incomes with which to purchase printed products, there is the problem of distribution. Roads along which delivery trucks might travel in Africa are in terrible condition, as Howard French reports from Zaire:

> [A European] diplomat, who recently toured each province, said he was shaken by what he had found.
>
> "Such roads as exist won't allow you to travel more than 15 miles an hour . . . to visit Equateur Province, even though it is the president's home, is to see things just as they were in the time of Stanley". . . . The highway to Kinshasha, which cuts across one grassy plateau after another before plunging through steep river-cut valleys, has deteriorated so badly that only a handful of interpid truckers take it anymore. . . . Fuel for the few cars in circulation is sold in whiskey bottles at the roadside, since there are no gas stations.[47]

Public bus and train service are unreliable or nonexistent in rural districts of many other countries, and shipments of magazines or newspapers sent in whatever vehicles are available risk being stolen, as do shipments sent by mail. Many of the "used" publications sold by African sidewalk vendors have actually been hijacked directly from buses or the post office, without ever having reached their original destinations.[48]

For the most part, Hachten reports:

> Newspaper readership has remained confined to the capitals and a few large cities where the few educated Africans, employed in either government or urban occupations, resided.
>
> Due to illiteracy, poverty, malnutrition, and linguistic diversity, the majority of Africans—peasants living on subsistence agriculture and many others crowded into urban slums—remain untouched by the printed word. Newspapers remain "European" in that they are elite institutions speaking to that minority who live in cities,

are educated, are literate, and generally run the government or are involved in the small modern sector. This is
obviously an important audience, but it is not the broad
public, and hence newspapers continue as an elite, not a
mass, medium.[49]

And, as the study by Olowu cited in chapter 1 has already
indicated, Africa's urban-based, independent print media—whether
they attempt to reach readers in the city or the countryside, or
both—do a poor job of reporting on agriculture and rural issues,
despite the need for such reports:

> The coverage pattern revealed by this investigation sug-
> gests that if Nigerian papers are to sustain campaigns on
> innovations, not only must the relative space allocated to
> agriculture increase, but a considerable increase in the
> coverage of special recommendations and marketing and
> pricing information is needed.[50]

On one hand, urban newspaper readers receive almost no
farm news. On the other, even when agriculture is covered, the
reports are of little use to rural people.[51] One possible reason for
this situation, ventured by Olowu, should be of particular interest to
journalism educators:

> Providing such utilitarian information [useful farm news]
> is predicated on the availability of resource persons with
> adequate knowledge of agriculture and journalistic prac-
> tice. It may be the paucity of resource persons with the
> appropriate mix of these qualities has a remote link to
> the scarcity of functionally relevant content in Nigerian
> newspapers.[52]

Of course, even the most authoritarian African governments realize that development programs aimed at improving farm
production cannot succeed if farmers themselves are left completely
out of the information loop. As Ochs observes:

> In the outlying areas are many who have never listened
> to radio regularly and have never seen a television set or
> a daily newspaper. To try to do something about this,

149

some nations have been hard at work for two decades to establish rural newspapers. Such efforts are an unpublicized feature of the African press scene. Evidently UNESCO is the only major organization that has taken an interest in them as a whole.[53]

Ochs errs here, overlooking the fact that "religion, in particular Christianity, has played a very big role in the development of the media in Africa."[54] Merrill notes that:

Mission newsletters and parish magazines containing church-related announcements have penetrated most rural and urban areas ever since missionaries first came to Africa in the second half of the 19th century (Martin 1991, p. 159). . . . Newspapers sponsored by Catholic and Protestant churches have continued to be published, serving the purpose of being not only extensions of the pulpit but also news informers and opinion leaders.[55]

However, the efforts of UNESCO, working with national governments, have been by far the most important in terms of scale and continuity. UNESCO has also been the most carefully detailed observer of press development in rural Africa. The overview provided by a 1980 UNESCO study is worth quoting at length, as most of its authors' observations remain true today:

Efforts to start rural newspapers in Africa began in the early 1960s in Liberia. These were usually one-page mimeographed news-sheets with a circulation of about 2,000. Almost all of them were linked to literacy projects.

As a method of mass production, the mimeograph technique offered a simple, low-cost entry for rural journalism. It did, however, have its disadvantages: limited reproduction and inability to reproduce illustrations and photos. The newer electronic copying machines could overcome the problems of limited reproduction, but at a higher cost and with more complicated problems of maintenance.

In order to solve these problems while at the same time keeping production costs within modest means, another strategy was used in the 1970s. Reporters and editors were trained in the villages ready to undertake a rural press project; simple editing and layout facilities were provided. Contracts were made with either a government or a commercial printer for the photographic plate-making and printing; the printed copies were then circulated in the project villages. The new methods permitted reproduction of photographs and illustrations, higher quality printing, and runs of up to 25,000, sometimes using four-color separation.

Evaluations of the progress of the rural press in Africa towards the beginning of 1977 showed two important points:

First, despite the success of the later generation of newspapers, and their role in literacy retention and the dissemination of development information, the total number and circulation of these papers have remained, at best, modest;

Second, side by side with the advent of the rural newspapers, more than a score of journalism training and research institutions have been established since 1965, but few of the rural journalists had formal training at any of these institutions nor did any of the newspapers benefit from their researches; and none of these institutions has a journalism course oriented directly to rural journalism.[56]

Following up the UNESCO study six years later, Ochs reported that "at latest count, 16 Black African countries had 53 rural newspapers published in at least 26 African languages." Despite their achievements, their negative aspects were obvious:

In all cases, these are government organs of "development journalism," promoting literacy, agriculture, public health or general economic development. A frankly stated objective of some is "popularization of government policies." The editors generally are government information officers. . . . Lack of training, low pay and poor working conditions are frequent handicaps.[57]

151

The 1980 UNESCO study noted that of the rural papers examined, "not one of them is economically independent or self-supporting. All are dependent on government subsidies for their existence."[58]

In addition, many "are used as a means of communicating government messages and are not newspapers with which members of the community can identify themselves."[59]

Of course, government-run publications, like government-run radio under authoritarian regimes, are unlikely to listen to farmers' criticisms of the government programs and projects they are assigned to promote. Grassroots feedback is rarely incorporated in the project documents. Nor do such media efforts, directed solely at rural audiences, reach the educated urban, policy-making or policy-influencing audience, who consequently remain ignorant of the effects of their policies in the countryside. The two-way information exchange crucial to the functioning of democracy is absent.

Besides UNESCO, a number of other outside agencies and NGOs have recognized these difficulties and attempted to help fill the rural coverage vacuum. For nearly 30 years, the FAO published *Ceres* magazine, a journal aimed at a broad readership of both lay and professional readers and benefiting from a mandate that permitted free and open debate on the issues. Printed in four languages, it covered agriculture, forestry, fisheries, and nutrition in a development context. FAO Director-General Jacques Diouf, a Senegalese national and former World Bank advisor who assumed his post in 1994, was reportedly angered by a November 1995 article[60] in the magazine that criticized some World Bank policies. He abruptly ordered the magazine "abolished" in early 1996.

Also suspended in the early 1990s were *Panoscope*, published by the London-based Panos Institute, and the award-winning *African Farmer*. Of the major foreign-based periodicals covering African agriculture, only *Afrique Agriculture*, published in Henonville, France, was still in existence in 1996.

Major African-based publications aimed at national farming readerships (as opposed to the local or district audiences served by the rural newspapers discussed above) include in-house membership publications like *Kenya Farmer* (circ. 25,000), published by the Agricultural Society of Kenya, and Namibia's *Agri Forum* (circ. 5,500), published by the Namibia Agricultural Union, as well as government periodicals such as *Agri-info* (circ. 2,000), published by the Rural Development Ministry of Chad, and *The Farmer* (circ. 600), published by the Gambian government.[61] Many such periodicals are aimed at the elite of the farming community who own large farms producing cash crops for export, as opposed to ordinary, small-scale subsistence farmers. For the most part, they make no attempt to reach urban audiences or to interpret farm news for city dwellers.

Government-owned radio, as already mentioned, is chronically cash-strapped, and rural radio is the poorest of African broadcasting's "poor cousins." Nor can rural stations hope to supplement their shrinking official budgets by "going commercial." As Jaya Weera of UNESCO's Division of Communication explains: "Urban areas are markets for advertising, while rural areas are not. Thus the very stations that need funds the most can't attract advertisers."[62]

The Canada-based Developing Countries Farm Radio Network (DCFRN), an NGO founded by former CBC farm broadcaster George Atkins, tries to provide some relief by promoting agricultural extension broadcasting in developing countries. As the network's Jennifer Pitte explains, however, the organization's efforts are confined largely to "sending out packages of technical farming information to broadcasters and extension workers, who use them in putting together their own programs."[63] With funds from the International Development Research Centre (IDRC), the NGO in 1992 helped set up an autonomous network based in Harare, Zimbabwe, to service East and Southern Africa. However, this network, like its parent organization, "does not have a station nor does it have broadcasters on its payroll. It writes scripts to answer a need."[64]

As for rural television, as Ochs observes, "educational television on a major scale does not exist" in Africa.[65]

In most of sub-Saharan Africa, neither the specialist agricultural media (aimed at educating and improving the lot of farmers), nor the general media (which should be educating the general population about key issues in agriculture), are sufficiently developed to do an effective job.

Journalism Training

Despite the many noxious effects of colonialism on African journalism, the earliest efforts at journalism education in the sub-Saharan region were instituted by Europeans—and the results were not wholly negative. As Hachten recounts:

> Historically, the best and most influential newspapers as models or prototypes for African journalism were those in the British colonies that mainly served the British settlers or commercial interests.
>
> The onetime Daily Mirror group papers—the *Daily Mail* in Freetown, the *Daily Graphic* in Accra, and the *Daily Times* in Lagos—had long since become government mouthpieces and propaganda sheets, but they still resembled, in form and style, their Fleet Street prototypes. Much the same impact was had by the East African Standard group of papers in Kenya, Tanzania and Uganda, as well as the Argus papers of Rhodesia/ Zimbabwe. These and other European papers provided a stimulating environment in which an African journalist could work and learn his trade. . . .
>
> The numerous journalism training courses—such as the IPI Training Scheme in Nairobi, those of the Thomson Foundation, and various others in the United States, Britain and France—reinforced these Western models of journalism. But it can be argued that, for West Africa certainly, the three London Mirror papers were a more influential and lasting school of journalism for African newsmen than the myriad training programs in and out of educational institutions.[66]

Among the lessons learned by early African journalists working on these British colonial papers were some that later backfired on the English settlers, especially after the Second World War. Merrill explains:

> The media, during this era, were more often than not willing partners in the ideological campaign to denigrate Africa and Africans by presenting them as primitive and as culturally and mentally inferior. The media of colonial Africa were used, among other reasons, to foster loyalty and conformity with the colonial system and to counter anticolonial and nationalist activities.

Still, the press during this period served as a critical force in mobilizing the indigenous population against colonial rule. African soldiers who had fought on the side of the Allies during World War II returned to Africa. Having been exposed to other cultures, they became aware of the inferior standard of living they were subjected to at home and began to question the idea of European superiority. Thus, the tradition of vociferous dissent emerged. Some editors and publishers started using their newspapers as a revolutionary tool in the liberation struggle. Many of them, notably Kwame Nkrumah of Ghana and Nnamdi Azikiwe of Nigeria, later became presidents of their countries.[67]

Ironically, the very press freedom that helped gain Africans their independence has since been suppressed in many African countries, as Hachten points out:

> The earlier generation of notable African journalists included the names of Peter Enahoro and Timothy Ulo Adebango of Nigeria, John Dumogo of Ghana, Kelvin Mlenga of Zambia, Hilary Ng'weno of Kenya, and Percy Qoboza of South Africa, to name just a few. As a mark of their excellence, all got into difficulties with their governments for editing outspoken independent newspapers. And the struggle between African journalists and their governments goes on.[68]

Journalism training programs in Africa today are centered on the journalism faculties of the continent's leading universities and technical schools, most of which are members of the Nairobi-headquartered African Council on Communication Education (ACCE).[69] Cooperating closely with representatives of UNESCO, the ACCE in 1988 produced a comprehensive directory of communication training institutions in Africa, describing their programs.[70] Representatives of several of these institutions were present at the 1991 International Meeting of Regional Communication Training Institutions for Communication Development, hosted by UNESCO in Paris, at which Africa's programs—along with those of other, non-African developing countries—were discussed.[71]

Many of the university Mass Communication programs listed in the ACCE Directory, particularly those located in Kenya and Nigeria, appeared wide-ranging and comprehensive, and several included training in "communication for rural development"—usually conceived as agricultural extension communications—among their activities. In Burkina Faso, for example, the Centre interafricain d'etudes en radio rurale de Ouagadougou (CIERRO) has as its chief aim "to train journalists in the use of communications, especially rural radio, to serve development purposes."[72] However, only three schools (as of 1995, the ACCE had 65 members) listed activities related to agricultural journalism, *per se*: (1) the Zambia Institute of Mass Communication, in Lusaka, noted that it publishes a *Rural Reporting Textbook*; (2) the University of Nairobi School of Journalism publishes a text entitled *Workshop on Rural Journalism*; (3) Tanzania Information Services, in conjunction with its journalism training activities, publishes *Tanzania Agriculture and Livestock*. Presumably, students enrolled in journalism training courses at these institutions have the opportunity to study from or contribute to these publications. None of the institutions listed any actual classroom training in farm journalism.

Participants at the 1991 Paris meeting at UNESCO cited a number of problems in African journalism education. Representatives of the Pan African News Agency (PANA) "complained about the dwindling support from UNESCO" (a result of budget cutbacks), while the U.S.-based Center for Foreign Journalists (CFJ) observed that "there seemed to have been little institutional growth in film training in Africa." Participants generally felt that communication research in Africa was "hampered by limited support from local institutions, dearth of funding for research, limited exchange of research information and a general lack of the culture of planning and evaluation." They also declared that "efforts should be supported to design research programs and instruments better related to communication activities in rural communities and grassroots situations."[73]

It should be noted as well that, while individual training programs may look good on paper, African schools on the ground—especially in the wake of World Bank/IMF structural adjustment programs that forced sharp cuts in educational funding—are often plagued by staff shortages, inadequate resources, and a lack of up-to-date textbooks and teaching materials. In addition, many training institutions—like most African broadcast stations and rural newspapers—are dominated by government. In Kenya, for example, the Kenya Institute of Mass Communication (listed as a training institution in the ACCE Directory) trains students in broadcast production "and government information work."[74] As Mwaura reports:

> Students for information work are recruited through the civil service and automatically become employees of the Ministry of Information and Broadcasting, which they join as information assistants upon graduation. . . .
>
> The training syllabus for information assistants is designed to produce information officers-cum-journalists. The students are expected to train in all aspects of the mass media—how to report and write for the press, radio and television as well as how to carry out the duties of a government information officer, which include government public relations.[75]

Today, the quality of sub-Saharan Africa's two most advanced journalism training systems—those of Nigeria and Kenya—may also suffer from the open, often violent hostility toward press freedom of dictators Sani Abacha and Daniel Arap Moi.

As in the case of farm-oriented publications, some external donors—notably the Scandinavian countries—have attempted to improve farm journalism training in Africa. In 1996, for instance, the Nordic-SADC (Southern African Development Community) Journalism Center in Maputo, Mozambique, supported by funds from the Danish aid agency DANIDA, offered a two-week course in "rural reporting" for working African journalists at Grahamstown, South Africa. The course was scheduled to be repeated in Tanzania.[76] The FAO has also published a highly praised textbook for rural radio broadcasters, based on African field experience.[77]

In 1995, representatives of the communications staffs of the FAO, the West African Riziculture Development Association (WARDA), the Netherlands-based Technical Center for Agricultural and Rural Cooperation (CTA), the International Council for Research in Agroforestry (ICRAF), and other organizations initiated discussions aimed at the organization of a series of training courses for African science writers and science journalists—including agricultural journalists. A few short workshops for science writers had already been conducted by CTA in Africa, and meeting participants were interested in a joint effort to expand the scope of these seminars. However, due to budget constraints, discussions remained preliminary.

A literature search, inquiries posted to the pertinent Internet discussion lists, and queries to members of the ACCE failed to identify any other training course in agricultural reporting in Africa.

Merrill states that:

> When looking at the current state of affairs as far as the training of journalists in Africa is concerned, it seems that this aspect desperately needs attention if African media are to become independent and to be judged according to the same standards as those in the rest of the world.[78]

In a region where farming is the chief occupation of two-thirds of the people, the situation would appear to be even more acute in regard to agricultural journalism.

CHAPTER NINE

Conclusions

The preceeding chapters have provided:

1. A survey of some of the current issues in agriculture, intended to demonstrate to journalists and journalism educators that the subject area is not only newsworthy, but crucial to the future of the human species itself.
2. A summary of both agricultural journalism practice, past and present, and the state of agricultural journalism training in three representative global regions, showing that both farm coverage and the resources available for training farm journalists are inadequate—in some cases, grossly inadequate.

Of necessity, the treatment of these subjects has been selective: an encyclopedic examination would have run to thousands of pages. Nevertheless, it is hoped that the examples given will alert both academics and members of the working press to the fact that a serious "blind spot" does, in fact, exist. One of the basics of civilization—of survival—is being neglected by precisely those whose role it is to serve as "society's immune response system,"[1] as the watchdogs who warn us of impending danger.

What should be done to remedy the situation is not the subject of this text. A thorough discussion of how present coverage could be improved, and what adjustments might be made to the curriculae of journalism training institutions so as to sensitize journalists to the importance of farm issues, should be the focus of further research. For example, as noted in chapter 6, a fruitful line of investigation might be to examine the statistical relationships between farm coverage, rural demographics, and advertiser preferences. Another area that merits research is curriculum development—particularly the development of training systems for working journalists from both the industrialized North and the Global South.

Outside North America, a number of limited experiments in training journalists to cover agriculture have been undertaken, most of them in the field of so-called "development journalism" in the Third World. For example, as already noted in chapter 8, the Nordic-SADC Journalism Centre, supported by grants from the Danish aid agency DANIDA, has conducted occasional short (usually two-week) courses for working African journalists. One course offered in 1996 at Grahamstown, South Africa, focused on "rural reporting," and was scheduled to be repeated in Tanzania. Personnel at a number of major development agencies, such as the participants from FAO, ICRAF, WARDA, and CTA mentioned in chapter 8, have also indicated interest in establishing short training courses for developing country writers specializing in science subjects, including agriculture. Efforts should be made to follow up on the interests of these development agencies, and their sometimes innovative experiments.

However, recent budget cutbacks prompted by the neo-conservative fiscal agendas of leading donor nations have severely curtailed hopes of branching into such activity. All over the Third World, training initiatives suffer from the twin handicaps of dwindling funds and lack of a clear donor mandate to focus on problems the donors themselves do not regard as having high priority.

In the northern industrial countries, meanwhile, farming has been largely abandoned as a subject of regular, consistent news coverage outside the specialized farm press. A few isolated attempts have been made by farm groups to catch the attention of mainstream society—and by extension to rekindle the interest of the mainstream media. For example, the Leopold Center for Sustainable Agriculture, in Ames, Iowa, describes two recent "consciousness-raising" projects:

> A coalition of about 25 Iowa ag industries and organizations have banded together to form the Coalition for Ag Image Promotion. After doing some research, the coali-

tion has focused much of its efforts at Iowa school-aged children, with the understanding that attitudes about ag and food are formed at early ages. They soon will unveil an ag resource library distributed to all grade school libraries and media centers, and have developed PSAs, *newspaper supplements* [emphasis added], and a summer teachers' academy that attracts 60 to 70 elementary teachers annually. Representatives of this group recently addressed a local chapter meeting of the Public Relations Society of America. . . .

The Iowa chapter of NAMA annually celebrates with an "almost free lunch," held in downtown Des Moines to attract this urban audience. A lunch of Iowa-produced foods is distributed for a donation of the typical earnings an Iowa producer realizes from his/her production. Proceeds are donated to the Food Bank of Iowa. Entertainment, ag quizzes, displays, *a live [radio] broadcast* [emphasis added], and friends of agriculture awards draw further interest.[2]

Of course, such small-scale local efforts at telling the farm story to the public are no substitute for adequate news coverage. The very fact that groups in a farm state like Iowa should feel compelled to go to such lengths to attract urban people's attention is itself a comment on the state of U.S. farm journalism. Perhaps journalism schools should be making an effort to seek out such groups, and develop ways to incorporate their messages in the curriculum.

But, whatever the solution, the first step in solving any problem is to identify its presence. Hopefully, this book has done so.

Notes

Chapter One

1. Joyce Egginton, *The Poisoning of Michigan* (New York: W.W. Norton & Co., 1980), 13.
2. *Ibid.*, 198–201.
3. *Editor & Publisher International Yearbook, 1975* and *1995* (New York: Editor & Publisher Co., Inc.)
4. *Broadcasting Yearbook 1976* and *Broadcasting & Cable Yearbook 1994* (New Providence, N.J.: R.R. Bowker).
5. Ann Reisner and Gerry Walter, "Agricultural journalists' assessments of print coverage of agricultural news," *Rural Sociology,* 59(3), 1994, 525–37.
6. *Ibid.*, 532.
7. Bill Doskoch, "Project Censored Canada: A top 10 list of underreported stories," *Media*, Fall 1995, 7.
8. Betty P. Brown, ed., RUSAG-l, "Russian ag communicator update." In FSUmedia, (fsumedia@sovam.com). 1 December 1995.
9. John C. Merrill, ed., *Global Journalism: Survey of International Communication* (White Plains, N.Y.: Longman Publishers USA, 1995), 182.
10. Paul Ansah et al., *Rural Journalism in Africa (Reports and papers on mass communication No. 88)* (Paris: UNESCO, 1981) 11.
11. Terry A. Olowu, "Reportage of agricultural news in Nigerian newspapers," *Journalism Quarterly*, 67(1), 1990, 195–200.
12. *Ibid.*, 199.
13. Association for Education in Journalism and Mass Communication (AEJMC), *Journalism and Mass Communication Directory 1995–1996* (Columbia, S.C.: AEJMC, 1996).

14. Viera Simkova. <trend@savba.savba.sk>. Personal e-mail, 4 October 1995.
15. Peter da Costa, <ipsdc@gn.apc.org>, Personal e-mail, 2 May 1995.

Chapter Two

1. Masanobu Fukuoka, *The One-Straw Revolution* (Emmaus, PA., Rodale Books Inc., 1978).
2. H.W. Fowler & F.G. Fowler, eds., *The Concise Oxford Dictionary of Current English*, 5th ed. (Oxford: Oxford University Press, 1964), 297.
3. Barbara Bender, *Farming in Prehistory: From Hunter-gatherer to Food-producer* (London: John Baker, 1975), 12.
4. Thomas Pawlick, "The essential ploughman," *Harrowsmith*, IV(7), 1980, 42–49, 104–106.
5. R.E. Allen, ed., *The Concise Oxford Dictionary of Current English*, 8th ed. (Oxford: Oxford University Press, 1992), 282.
6. John Steinbeck, *East of Eden* (London: Penguin Books Ltd., 1979).
7. Raymond Williams, *The Country and the City* (London: The Hogarth Press, 1993), 1.
8. *Ibid.*, 1.
9. Nino Ricci, *Lives of the Saints* (Toronto: McKay, 1991).
10. Carol Harrington, "Calgary cancels cowboy festival due to lack of interest," *Canadian Press*, 4 July 2000.
11. Detailed descriptions of this aspect of African history are contained in: Elspeth Huxley, *White Man's Country* (London: Chatto & Windus, 1935); and Errol Trzebinski, *The Kenya Pioneers* (London: William Heinemann Ltd., 1985).
12. Williams, *op. cit.*, 291.
13. *Ibid.*
14. *Ibid.*, 292–93, 297.
15. Murray Bookchin, "Radical Agriculture," in *Radical Agriculture*, ed. Richard Merrill (New York: New York University Press, 1976), 3–13.
16. Thomas Pawlick, "The cause and its effects," *Harrowsmith*, VII(2), 1982, 29.
17. FAO, *The State of Food and Agriculture* (SOFA) 1994 (Rome: FAO, 1994).
18. A good description of the situation in Third World cities is provided in: Jorge Hardoy, Diana Mitlin & David Satterthwaite,

Environmental Problems in Third World Cities (London: Earthscan Publications Ltd., 1993).

19. FAO, *op. cit.*
20. Nick Kotz, "Agribusiness," in *Radical Agriculture*, ed. Richard Merrill (New York: New York University Press, 1976), 41–51.
21. *Ibid.*, 48.
22. *Ibid.*, 49.
23. Bookchin, *op. cit.*, 4.
24. Wendell Berry, *The Unsettling of America* (San Francisco: Sierra Club Books, 1977), 40.
25. *Ibid.*, 41.
26. Don Lajoie, "Proposed greenhouse big as four farms," *The Windsor Star,* 23 June 2000, A1.
27. Berry, *op. cit.*, 43.
28. *Ibid.*, 47.
29. *Ibid.*, 9.
30. Louise Elliott, "Crack on rise in rural Canada, experts warn: More young people using," *Canadian Press,* 23 July 2000.
31. Berry, *op. cit.*, 6.

Chapter Three

1. Riad El-Ghonemy and Jeremy Avis, "Food security=social security," *Ceres,* 27(2), 17.
2. FAO, *The State of Food and Agriculture (SOFA) 1995* (Rome: FAO, 1995), 244.
3. *Loc. cit.*
4. *Ibid.*, 200.
5. *Ibid.*, 245, 247.
6. *Ibid.*, 247.
7. Stefan Tangermann, "A major step in a good direction," *Ceres,* 27(1), 24–27.
8. *Ibid.*, 26.
9. *Loc. cit.*
10. Tim Lang and Colin Hines, *The New Protectionism* (London: Earthscan Publications Ltd., 1993).
11. Tim Lang and Colin Hines, "A disaster for the environment, rural economies, food quality and food security," *Ceres,* 27(1), 19–23.
12. Terry Pugh, "Thousands of family farmers will become casualties to this adjustment," *Ceres,* 27(1), 1995, 28–33.
13. For a detailed description of the appointment process and its effect on World Bank/IMF policy, see Susan George and Fabrizio

Sabelli, *Faith and Credit: The World Bank's Secular Empire* (London: Penguin Books, 1994).

14. Pugh, *op. cit.*, 31.
15. *Canadian Press*, "U.S. gears up for food fight with Canada," *The Ottawa Citizen* 30 January 1996, A7; and Drew Fagan, "Canada defends farm tariffs," *The Globe and Mail*, 30 April 1996, A1–2.
16. Thomas Pawlick, "The cause and its effects," *Harrowsmith*, IV(7), 1982, 29.
17. *Ibid.*, 28.
18. Pugh, *op. cit.*, 29.
19. *Ibid.*, 30.
20. A search by the author found that most Canadian coverage of the GATT's effects on farming (with the exception of that of the *Globe and Mail*, whose columnist David Roberts has been able to provide occasional glimpses of farm life as it is actually lived) tends to focus on the macro-economic trade picture, rather than on individual communities. Much of it consists of wire copy reports originating in the United States.
21. Linda Turk, "No sacred cows in the NAFTA era," *The Globe and Mail*, 25 April 1996, A20.
22. *Ibid.*
23. Drew Fagan, *op. cit.*
24. Dina Temple-Raston, "Family farms can't beat corporate muck," *USA Today*, 23 May 2000 (reprinted in *The Detroit News*, 24 May 2000, 2A).
25. Raymond Williams, *The Country and the City* (London: The Hogarth Press, 1993), 291.
26. Immanuel Wallerstein, *The Modern World System I: Capitalist Agriculture and the Origins of the European World-economy in the Sixteenth Century* (San Diego, Calif.: Academic Press Inc., 1974).
27. Lang and Hines, "A disaster for the environment," 21.
28. Riad El-Ghonemy, *The Dynamics of Rural Poverty* (Rome, FAO, 1986).
29. El-Ghonemy and Avis, *op. cit.*, 17.
30. *Ibid.*, 18.
31. Lang and Hines, "A disaster for the environment," 20.
32. *Op. cit.*, 203–204.
33. *Loc. cit.*
34. "Future directions for Canadian agriculture and agri-food" (Ottawa, Agriculture and Agri-food Canada), September 29, 1994, 19–19A.
35. Clifford Cobb, Ted Halstead, and Jonathan Rowe, "If the GDP is up, why is America down?" *The Atlantic Monthly*, October 1995, 36.

36. John Ibbitson, "The high cost of Common Sense," *The Globe and Mail,* 24 May 2000, A1.
37. FAO, The State of food and agriculture (SOFA 94), Rome: United Nations Food and Agriculture Organization, 1994.

Chapter Four

1. Masanobu Fukuoka, *The Natural Way of Farming* (New York: Japan Publications, Inc., 1985), 35.
2. Under Canada's "supply-management" farm marketing system, commodity marketing boards, such as the Ontario Milk Marketing Board or the Canadian Egg Marketing Agency, grant farmer/members "licenses"—referred to as quota—to sell specific amounts of a given board-regulated farm product. Quota can be bought and sold, and frequently changes hands when a farmer dies or retires.
3. Gordon R. Conway and Jules N. Pretty, *Unwelcome Harvest: Agriculture and Pollution* (London: Earthscan Publications, 1991), 1.
4. Mike Collinson, "Green evolution," *Ceres,* 27(4), 23–24.
5. *Ibid.,* 24.
6. Conway and Pretty, *op. cit.,* 2.
7. *Ibid.,* 7.
8. Epidemiological statistics from Vietnam, and detailed descriptions of the physiochemical processes by which dioxin affects living organisms, can be examined in: Arthur H. Westing, *Herbicides In War: The Long-term Ecological and Human Consequences* (New York: Peace Studies, 1984).
9. Conway and Pretty, *op. cit.* 45.
10. *Ibid.,* 37.
11. James Hughes, ed., *The Larousse Desk Reference* (New York: Larousse Kingfisher Chambers Inc., 1995), 81.
12. Conway and Pretty, *op. cit.,* 27.
13. R. Repetto, *Paying the Price: Pesticide Subsidies in Developing Countries* (Washington: World Resources Institute, 1985).
14. K.D. Switzer-Howse and D.R. Coote, "Agricultural practices and environmental conservation," Ottawa: Agriculture Canada, 1984, 12.
15. Conway and Pretty, *op. cit.,* 198.
16. Switzer-Howse and Coote, *op. cit.,* 12.
17. E. Paul Taiganides, "The animal waste disposal problem," *Agriculture and the Quality of Our Environment,* ed. Nyle C. Brady (Washington, D.C.: American Association for the Advancement of Science, 1967) 389–90.

18. *Loc. cit.*
19. Switzer-Howse and Coote, *op. cit.*, 13.
20. Natalie James, "Report arouses concern over government role in E. coli deaths," *The Canadian Press,* 28 July 2000.
21. Tom Spears, "Swill from the swine," *The Windsor Star,* 10 June 2000: G1.
22. For an in-depth discussion of these patterns and their effects on the environment, see Lennart Hansson, Lenore Fahrig and Gray Merriam, eds., *Mosaic Landscapes and Ecological Processes* (New York: Chapman and Hall, 1995).
23. Raymond J. O'Connor and Michael Shrubb, *Farming and Birds* (Cambridge: Cambridge University Press, 1986), 80.
24. *Ibid.,* 149.
25. *Ibid.,* 83.
26. *Ibid.,* 83–84.
27. *Loc. cit.*
28. *Ibid.,* 86.
29. Stewart Elgie, "Bite: Endangered species need a law that gives them some," *Environment Views,* 18(1), 21.
30. David Wylynko, "The rate debate: Will the end of the transportation subsidy for prairie wheat lead to more sustainable farming practices on the Great Plains?" *Nature Canada,* 25(1), 17–21.
31. Kevin Van Tighem, "Save the gopher," *Environment Views,* 18(1), 16–19.
32. Edward B. Barbier et al., *Elephants, Economics and Ivory* (London: Earthscan Publications Ltd., 1990), 14.
33. *Ibid.,* 17.
34. Dick Pitman, "Wildlife as a crop," *Ceres,* 22(1), 30.
35. *Ibid.,* 30–35.
36. *Ibid.,* 35.
37. R.B. Martin, "A voice in the wilderness," *Ceres,* 26(6), 26, 27.
38. *Ibid.,* 26.
39. *Loc. cit.*
40. Switzer-Howse and Coote, *op. cit.*, 22.
41. *Loc. cit.*
42. *Ibid.,* 8.
43. *Ibid.,* 23.
44. *Ibid.,* 24.
45. Sandra Postel, "Waters of strife," *Ceres,* 27(6), 1995, 19.
46. *Ibid.,* 20.
47. *Ibid.,* 21.
48. James Hughes, *op. cit.*, 146.
49. Postel, *op. cit.*, 23.

50. B. Appelgren and S. Burchi, "The Danube's blues," *Ceres*, 27(6), 24–28.
51. Armelle Braun, "The megaproject of Mesopotamia," *Ceres*, 26(2), 25–30.
52. Polly Stroud, "Africa's wave of the future, or a backwash from the past?" *Ceres*. 27(4), 37.
53. Patricia Baeza-Lopez, "A new plant disease: Uniformity," *Ceres*, 26(6), 1994, 41.
54. Pawlick, "The cause and its effects," *op. cit.*, 35.
55. *Loc. cit.*
56. Baeza-Lopez, *op. cit.*, 44.
57. Vandana Shiva, "Mistaken miracles," *Ceres*, 27(4), 28–29.
58. *Ibid.*, 29–30.
59. Stroud, *op. cit.*, 39.
60. John Herbert, "The narrowing of the options," *Ceres*, 26(6), 1994, 42–45.
61. Rural Advancement Foundation International <http://www.rafi.org>, "USDA refuses to abandon Terminator technology," 28 July 2000.
62. Rural Advancement Foundation International <http://www.rafi.org>, "Earmarked for extinction? Seminis eliminates 2,000 varieties," 21 July 2000.
63. *Loc cit.*
64. Rebecca Goldburg, "Pause at the amber light," *Ceres*, 27(3), 1995, 21.
65. Alfred W. Crosby, *Ecological Imperialism: The Biological Expansion of Europe, 900–1900* (Cambridge: Cambridge University Press, 1986), 75.
66. James Hughes, *Larousse, op.cit.*, 126, 136.
67. Crosby, *op. cit.*, 154–55.
68. Goldburg, *op. cit.*, 23.
69. Rick Weiss, "Insect Bambi threatened by gene-altered corn," *The Ottawa Citizen*, 20 May 1999, A13.
70. James Walston, "C.O.D.E.X. spells controversy," *Ceres*, 24(4), 28–29.
71. Turning Point Project, "Unlabelled, untested . . . and you're eating it," advertisement No. 2 in a series on genetic engineering, 14 October 1999.

Chapter Five

1. Bill Mollison, quoted in John Madeley, "A design science with an ethic," *Ceres*, 24(6), 27.

2. *Op. cit.*, 25–27.
3. Vandana Shiva, "Mistaken miracles," *Ceres*, 27(4), 28, 31.
4. Helen Gillman and Helen Grimaux, "Utopia as judgement: The agricultural visionary's dream is also a criticism—one we can't afford to ignore," *Ceres*, 24(6), 1992, 16.
5. *Ibid.*, 19.
6. Madeley, *op. cit.*, 25.
7. Stephanie Power, "Feds shut scientist down," *Capital City*, 16–22 July 1998, 7.
8. A point source is a location from which pollutants eminate, which is sufficiently distant from the areas it affects, compared to its length and breadth, to be considered a point—for example, a single farm that pollutes a neighboring stream. A non-point source is a more general location—all of the farms in a river valley, for example, whose pollutants together might affect an entire watershed.
9. Committee on the Role of Alternative Farming Methods in Modern Production Agriculture, Board on Agriculture, National Research Council, *Alternative Agriculture* (Washington, D.C.: National Academy Press, 1989).
10. Gillman and Grimaux, *op. cit.*, 19.
11. ILEIA, Center for Research and Information Exchange in Ecologically Sound Agriculture, *Working for Ecologically Sound Agriculture* (Leusden: The Netherlands Ministry of Development Cooperation, 1996), 1.
12. Miriam Bianco, "No more finger in the dike: Holland turns back history's tide," *Ceres*, 26(1), 1994, 12–13.
13. For a detailed report on the Dutch plan, see Alex Steffen and Alan Atkisson, "The Netherlands' radical, practical Green Plan," *Whole Earth Review*, No. 87, Fall 1995, 94–99.
14. Shane Cave, "Rethinking a way of life," *Ceres*, 27(2), 38–40.

Chapter Six

1. Anecdotal evidence indicates that conditions described in this chapter for North America are similar in other northern industrialized countries, especially those of Western Europe.

   For example, France is still one of the most rural nations of the European Union (EU), and the French government's defense of its farm sector was one of the most controversial aspects of the recent Uruguay Round of the GATT negotiations. However, a review of the 1995 membership roll of the French *Association des Journalistes Scientifiques de la Presse d'Information*—whose "main aim is to promote cooperation among researchers from all disciplines, with a view to ensuring responsible and objective reporting of scientific issues to the public"—shows the following coverage specialties listed: life sciences, 28; medicine, 25; technology, 27; environment, 15; generalist, 52; politics of science, 1; astronomy/space, 18; physics/chemistry, 16; psychology, 1; archeology, 9; earth sciences, 7; paleontology, 1; mathematics, 1; agriculture, 0.

Unfortunately, however, in France and other EU countries there is no equivalent to the North American *Editor & Publisher International Yearbook*, with its detailed listings of daily newspaper staffs by subject specialty, and thus it is not possible to easily trace recent staffing patterns outside the specialized farm press.

Accordingly, this chapter focuses on North America, where available industry sources permit the assembly of a more complete picture.

2. Max Armstrong. <maxarm@aol.com>. WGN, U.S. Farm Report, Chicago. Personal e-mail, 31 October 1995.

3. James Romahn. <jromahn@web.apc.org>. Personal e-mail, 15 March 1996.

4. Michael Strathdee. <strathdeem@kwrmsnt.cmail.southam.ca>. Personal e-mail, 14 March 1996.

5. John Neil. <crs0148@inforamp.net>. Personal e-mail, 15 March 1996.

6. Romahn, *Loc. cit.*

7. Clifton Anderson, "Farm issues and the media," *Editor & Publisher*, 121(49), 56.

8. Paul Queck, personal communication, 10 November 1995.

9. Peter Hendry. <phendry@uoguelph.ca>. Personal e-mail, 19 October 1995.

10. Communications Branch, Agriculture and Agri-Food Canada, Product evaluation: Final report, Summer 1994, 4–5.

11. Brigid Rivoire, Agriculture and Agri-Food Canada, personal notes from panel discussion "Agri-Food: The forgotten beat," Ottawa, 29 June 1993.

12. Ron Milito, Agriculture and Agri-Food Canada, personal communication, 18 October 1995.

13. Armstrong, *op. cit.*

14. Neil Morton, "Why is the *Globe and Mail* selling us the wrong environmental story?" *Ryerson Review of Journalism*, Summer 1995, 13.

15. Doskoch, *op. cit.*

16. Wayne F. Alda. <wfalda@aol.com>. Personal e-mail, 25 March 1996.

17. Tana Kappel. <apbtk@msu.oscs.montana.edu>. Personal e-mail, 25 March 1996.

18. Michael Balter, "Does anyone get GATT?" *Columbia Journalism Review*, May-June 1993, 46–49.
19. Bob Davis. <ift@soli.inav.net>. Personal e-mail, 30 November 1995.
20. T. Joseph Scanlon, "A study of the contents of 30 Canadian daily newspapers for Special Senate Committee on Mass Media," Carleton University, 31 October 1969, 7.
21. *Editor & Publisher International Yearbook, 1995* (New York: Editor & Publisher Co., Inc.), I-1–I-449.
22. *Ibid.*, III-1–III-25.
23. *Editor & Publisher International Yearbook, 1999* (New York: Editor & Publisher Co., Inc.), I-1–I-499.
24. Ibid., I-500–I-527.
25. *Editor & Publisher International Yearbook, 1955* (New York: Editor & Publisher Co., Inc.), 20.
26. *Editor & Publisher International Yearbook, 1995*, (New York: Editor & Publisher Co., Inc.), 9.
27. *Op. cit.*, 1955, 236–40.
28. *Editor & Publisher International Yearbook, 1975*, (New York: Editor & Publisher Co., Inc.), 282–97.
29. *Op. cit.*, 1995, 9.
30. *Op. cit.*, 1955, 20, 193–94.
31. *Op. cit.*, 1975, 7, 97–102.
32. *Op. cit.*, 9, I-139–I-149.
33. Milito, *op. cit.*
34. Kevin Cavanagh, personal interview, 28 June 1996.
35. Hendry, *op. cit.*
36. Michael Cooke, personal interview, 28 June 1996.
37. A.J. Blauer, "Down the tubes: The dwindling future of private radio news," *Ryerson Review of Journalism*, Summer 1995, 10.
38. William Lazer, *Handbook of Demographics for Marketing and Advertising* (New York: Lexington Books, 1994), 96.
39. *Ibid.*, 99.
40. *Loc. cit.*
41. *Ibid.*, 241–42.
42. Statistical Service Unit, Policy Analysis Branch, 1993 Agricultural Statistics for Ontario (Toronto: Ontario Ministry of Agriculture, Food and Rural Affairs, August 1994), 12.
43. Lazer, *op. cit.*, 39.
44. *Op. cit.*, 6.
45. C. Edwin Baker, *Advertising and a Democratic Press* (Princeton, N.J.: Princeton University Press, 1994), 17.
46. *Ibid.*, 25–26.
47. *Ibid.*, 62–63.

48. *Ibid.*, 65–66.
49. *Ibid.*, 64.
50. *Ibid.*, 67.
51. *Ibid.*, 66–67.
52. Mark Fitzgerald, "A year of turmoil," *Editor & Publisher*, 129(1), 9.
53. As Reisner and Walter noted ("Agricultural journalists' assessments of print coverage of agricultural news," *Rural Sociology*, 59(3), 1994, 527):

> the criticisms of agricultural magazines indicate that their content for the same reason may tend to reflect the values and positions of agricultural (rather than general-public) interests. Critics of agricultural magazines argue that reporters for such magazines identify too closely with agriculture and take a pro-industry point of view, saying that farm magazine reporters write stories that are actually advertisements for companies or agribusiness in general. The critics also say farm magazine writers overemphasize higher production as a solution for agricultural problems, write too few stories about social issues in agriculture, ignore environmental problems created by current agricultural practices, and inadequately investigate potential scandals within agriculture. This situation is likely enhanced by the magazines' dependence on agribusiness advertisers and the reporters' own agricultural backgrounds.

54. Baker, *op. cit.*, 69, citing William Blankenburg.
55. Hendry, *op. cit.*
56. Hugh Owen. <owen@ornet.or.uoguelph.ca>. Personal e-mail, 28 October 1995.
57. "Guardians of a way of life: a fresh perspective on marketing boards," *Harrowsmith*, VII(2), 21–43, 108–109.
58. Anita Lahey, "Selling the farm: *Harrowsmith's* slide from green power to Green Acres," *Ryerson Review of Journalism*, Summer 1993, 33.
59. Reisner and Walter, *op. cit.*, 526.
60. William B. Ward, *Reporting Agriculture through Newspapers, Magazines, Radio, Television* (Ithaca, N.Y.: Comstock Publishing, 1959), 26–27.
61. Susan Bourette, "Borrowed helicopter helps police nab four," *The Globe and Mail*, 30 June 2000, A15.
62. "Nunavut files court challenge to claim Inuit exemption from federal gun law," *Canadian Press*, 22 June 2000.
63. Nor is it in France, despite the political and cultural importance of agriculture to the nation. A survey by this author of the leading French schools of journalism in 1995 found none with a specific

program in farm journalism. The Centre de Formation et de Perfectionnement des Journalistes, in Paris, offered occasional seminars on agriculture for its students, but no regular courses. The Ecole Superieure de Journalisme, also in Paris, offered a single course in "Third World political and economic problems," which touched only briefly on agriculture.

Only the Ecole Superieure de Journalisme de Lille regarded agriculture as a subject worth special attention. Its director, Patrick Pepin, noted that "we are in the process of creating a farm journalism stream, for 1996 or 1997. It will be open to agricultural engineers and students with science and technical masters degrees. It will involve a partnership with the Institut superieur d'Agriculture de Lille."

In Great Britain, only one farm journalism course could be located, a weeklong seminar sponsored jointly by the Guild of Agricultural Journalists and John Deere Ltd. (farm machinery manufacturers). The brief course was "aimed at agricultural students keen to get jobs in the agricultural and horticultural media," rather than in the mainstream press.

64. AEJMC Directory, *op. cit.*

65. Linda Billings. <billings@anamorphosis.usra.edu>. Personal e-mail, 25 September 1995.

66. Professor Lamar W. Bridges, Department of Journalism and Printing, East Texas State University, personal communication, 22 January 1996.

67. Universities offering journalism courses include Carleton University, Concordia University, University of King's College, Universite Laval, University of Regina, Ryerson Polytechnic University, Universite de Sherbrooke, Simon Fraser University, and University of Western Ontario. Community colleges offering journalism courses include Algonquin College, Centennial College, Fanshawe College, Holland College, Humber College, Loyalist College (in cooperation with Kemptville College), Niagara College of Applied Arts and Technology, Red River Community College, Seneca College and Westviking College.

68. Ann Reisner, "Course work offered in agricultural communications programs," *Journal of Applied Communications*, 74(1), 1990, 18–25.

69. Owen Roberts. <owen@ornet.or.uoguelph.ca>. Personal e-mail, 18 September 1995.

70. Course prospectus, "Agricultural Journalism Department: scientific, technical and agricultural communication," Department of Agricultural Journalism, University of Wisconsin-Madison, 1995.

71. Reisner, *op. cit.*, 23.

Chapter Seven

1. Esther Kingston-Mann, "Breaking the silence: an introduction," in *Peasant Economy, Culture and Politics of European Russia, 1800–1921*, eds. Esther Kingston-Mann and Timothy Mixter (Princeton, N.J.: Princeton University Press, 1991), 3–4.
2. Kingston-Mann, *op. cit.*, 4.
3. Emmanuel Wallerstein, *The Modern World System I* (San Diego, Calif.: Academic Press Inc., 1974), 252.
4. Robert B. Costello, ed., *Random House Webster's College Dictionary*, 3d ed. (New York: Random House, 1995), 1546.
5. Kingston-Mann, *op. cit.*, 5.
6. *Ibid.*, 6–7.
7. Jeffrey Burds, "The social control of peasant labor in Russia: The response of village communities to labor migration in the Central Industrial Region, 1861–1905," in *Peasant Economy, Culture and Politics of European Russia, 1800–1921*, eds. Esther Kingston-Mann and Timothy Mixter (Princeton, N.J.: Princeton University Press, 1991), 54–55.
8. Burds, *op. cit.*, 55.
9. James H. Krukones, *To the people: The Russian government and the newspaper Sel'skii Vestnik (Village Herald), 1881–1917* (New York: Garland, 1987).
10. *Ibid.*, 253, 256.
11. David Christian, "The black and gold seals: Popular protests against the liquor trade on the eve of emancipation," in *Peasant Economy, Culture and Politics of European Russia, 1800–1921*, eds. Esther Kingston-Mann and Timothy Mixter (Princeton, N.J.: Princeton University Press, 1991), 270.
12. Christian, *op. cit.*, 271.
13. *Loc. cit.*
14. Orlando Figes, "Peasant farmers and the minority groups of rural society: peasant egalitarianism and village social relations during the Russian Revolution (1917–1921)," In *Peasant Economy, Culture and Politics of European Russia, 1800–1921*, eds. Esther Kingston-Mann and Timothy Mixter (Princeton, N.J.: Princeton University Press, 1991), 386.
15. Scott Shane, *Dismantling Utopia: How Information Ended the Soviet Union* (Chicago: Ivan R. Dee, 1994), 87–88.
16. *Loc. cit.*
17. *Loc. cit.*
18. T.L. Gettings and George DeVault, "The Russian revelation," *The New Farm*, 17(4), 34.
19. John Murray, *The Russian Press from Brezhnev to Yeltsin: Behind*

the Paper Curtain (Dublin, Ireland: Trinity College Press, 1994), 27–28.

20. *Loc. cit.*
21. *Loc. cit.*
22. *Loc. cit.*
23. *Op. cit.*, 120–21.
24. *Loc. cit.*
25. *Op. cit.*, 123.
26. *Loc. cit.*
27. Figes, *op. cit.*
28. Shane, *op. cit.*, 98.
29. Murray, *op. cit.*, 126.
30. *Ibid.*, 29.
31. Kim Carter and Scott McClellan, "Time is no friend to Poland's peasant farmers," *Ceres*, 23(2), 1991, 21.
32. Adrian Civici, Donika Kercini and Michael Griffin, "A nation of postage stamps," *Ceres*, 26(2), 1994, 40.
33. Carter and McClellan, *op. cit.*, 25–26.
34. *Ibid.*, 27.
35. Civici, Kercini, and Griffin, *op. cit.*, 40.
36. *Loc. cit.*
37. Stephen K. Wegren, "Rural migration and agrarian reform in Russia: A research note," *Europe-Asia Studies*, 47(5), 877–88.
38. *Ibid.*, 881.
39. John B. Dunlop, "Will a large-scale migration of Russians to the Russian Republic take place over the current decade?" *International Migration Review*, 27(3), 605–37, and "Will the Russians return from the near-abroad?" *Post-Soviet Geography*, 35(4), 204–15.
40. *Ibid.*, 206.
41. *Ibid.*, 214.
42. *Ibid.*
43. R. Ryvkina and R. Turovskiy, *The Refugee Crisis in Russia* (Toronto: York Lanes Press, 1993), 6.
44. *Ibid.*, 24–25.
45. Wegren, *op. cit.*, 879.
46. *Ibid.*, 884.
47. *Ibid.*, 883.
48. *Ibid.*, 885.
49. For detailed descriptions of the state of CIS farm "privatization," see: Peter R. Craumer, "Regional patterns of agricultural reform in Russia," *Post-Soviet Geography*, 35(6), 1994, 329–51; A. Petrikov, "Rural society and agrarian reform," *Problems of Economic Transition*, 37(6), 1994, 19–28; Stephen K. Wegren,

"Dilemmas of agrarian reform in the Soviet Union," *Soviet Studies*, 44(1), 1992, 3–36.

50. Carter and McClellan, *op. cit.*, 25.
51. Shane, *op. cit.*, 98.
52. Robert T. McMillan, "Some observations on Oklahoma population movements since 1930," *Rural Sociology*, 1(3), 1935, 334, 339.
53. *Op. cit.*, 340.
54. RUSAG-L: Current Events #79 <RUSAG-L@UMDD.UMD.EDU>, 8 July 1996, 1–8.
55. RUSAG-L: Current Events #73 <RUSAG-L@UMDD.UMD.EDU>, 15 February 1996.
56. *Op. cit.*, 9–11.
57. Editor <editor@uu6.psi.com>, 22 November 1995.
58. OMRI Daily Digest <fsumedia@sovam.com>, 24 November 1995.
59. From Moscow *Times* <fsumedia@sovam.com>, 15 October 1995.
60. From International Freedom of Expression Exchange Clearinghouse <fsumedia@sovam.com>, 17 October 1995.
61. From Guiragos Manoyan <fsumedia@sovam.com>, 30 September 1995.
62. From OMRI This Week <fsumedia@sovam.com>, 2 December 1995.
63. From Eric Johnson, Internews <fsumedia@sovam.com>, 6 July 1995.
64. *Reuters Limited*, "Police search for Russian cameraman," reported by fsumedia <fsumedia@sovam.com>, 11 July 2000.
65. Nick Wadhams, "Russia said restricting media," *The Associated Press*, 1 August 2000.
66. Ben Arus and Marcus Warren, "Media chief arrested in press purge by Kremlin," *The Electronic Telegraph*, 14 June 2000.
67. *Loc. cit.*
68. From Nicholas Pilugin, USIA Media Assistance Clearinghouse <nwpilugin@glas.apc.org>, reported on FSUMedia <fsumedia@sovam.com>, 17 July 1995.
69. OMRI Daily Digest, reported by Laurie Belin (belin@omri.cz) on FSUMedia <fsumedia@sovam.com>, 6 July 1995.
70. OMRI Daily Digest <fsumedia@sovam.com>, 11 July 1995.
71. Yuri Rokityanski, "TV should be produced not only in metropolis," International Media Centre/Internews <rockit@sovam.com>, reported on FSUMedia <fsumedia@sovam.com>, 6 July 1995.
72. International Labor Organization (ILO), "Trouble in the transition," *World of Work*, No. 12, 1995, 31.

73. Rokityanski, *op. cit.*
74. Simkova, <trend@savba.savba.sk>. Personal e-mail, 4 October 1992.
75. Nicholas Pilugin. <nwpilugin@glas.apc.org>. USIA Media Assistance Clearinghouse. Personal e-mail, 6 February 1996.
76. Hilmi Toros, Chief, Press Section, United Nations Food and Agriculture Organization (FAO), personal interview, 17 June 1996.
77. Gettings and DeVault, *op. cit.*, 34.
78. Agricultural Communicators in Education (ACE) <ace@gnv.ifas.ufl.edu>, reported on FSUMedia <fsumedia@sovam.com>, 8 December 1995.
79. Association for Education in Journalism and Mass Communication (AEJMC), *Journalism and Mass Communication Directory 1994–1995* (Columbia, S.C.: AEJMC, 1995), 73–88.
80. Frank A. Aycock, Associate Professor of Journalism, American University in Bulgaria, personal communication, 30 April 1995.
81. Center for Foreign Journalists: CFJ Clearinghouse, reported on East European Media List <eemedia@mcfeeley.cc.utexas.edu>, 14 July 1995.
82. *Ibid.*
83. Jan Bierhoff. <bierhoff@ejc.nl>. Managing director, European Journalism Center. Personal e-mail, 12 July 1995.
84. USIA Media Assistance Clearinghouse, reported on FSUMedia List <fsumedia@sovam.com>, 5 July 1995.
85. Jay Brodell. <brodellj@mscd.edu>. Personal e-mail, 24 April 1995.
86. From International Journalists' Network (IJNet), "CEE media organizations appeal for international support," 10 July 2000, reported on MIDEUR-L <mideur-l@listserv.acsu.buffalo.edu>.

Chapter Eight

1. William A. Hachten, *The Growth of Media in The Third World: African Failures, Asian Successes* (Ames, Ia.: Iowa State University Press, 1993), 54.
2. The author lived and worked in Kenya during 1989 and 1990, and the description of VOK news broadcasts is taken from actual programs taped during the period.
3. Timothy Appleby, "Rising stability puts Costa Rica at crossroads," *The Globe and Mail*, 6 July 1996: A6.
4. John Merrill, *op. cit.*, 209.
5. *Loc. cit.*

6. FAO, *The State of Food and Agriculture (SOFA95)*, Rome: United Nations Food and Agriculture Organization, 1995, 75.

7. John Merrill, *op. cit.*, 210.

8. FAO, SOFA95, *op. cit.*, 77.

9. *Ibid.*, 76.

10. *Ibid.*, 78.

11. Gurirab, T.B., as cited in John Merrill, *op. cit.*, 211.

12. Angelique Haugerud, *The Culture of Politics in Modern Kenya* (Cambridge: Cambridge University Press, 1995), 41–44.

13. *Loc. cit.*

14. David Blair, "Land grab in final phase," *The Daily Telegraph*, 14 July 2000, reprinted in *The Windsor Star*, 15 July 2000, A15.

15. John Merrill, *op. cit.*, 214.

16. *Ibid.*, 217.

17. Peter Mwaura, *Communication Policies in Kenya* (Paris: UNESCO, 1982), 54.

18. *Loc. cit.*

19. Hachten, *op. cit.*, 38.

20. John Merrill, *op. cit.*, 225–26.

21. *Ibid.*, 252.

22. *Ibid.*, 226.

23. Mwaura, *op. cit.*, 57.

24. Hachten, *op. cit.*, 38.

25. Hughes, *op. cit.*, 255.

26. *Loc. cit.*

27. Martin Ochs, *The African Press* (Cairo, Egypt: The American University in Cairo, 1986), 47–48.

28. *Loc. cit.*

29. Hughes, *op. cit.*, 254.

30. *Loc. cit.*

31. Ochs, *op. cit.*, 48.

32. Despite its title, which would indicate an audience made up exclusively of lawyers, the *Monthly* contained much political reporting and was widely read by educated Kenyans who were not themselves members of the legal profession.

33. These are the cover prices of issues of these magazines purchased by the author at newsstands in Nairobi in September 1992. The exchange rate at the time was approximately 30 Ks to US$1.

34. Patrick Nagle, Southam News, "Kenya: Tragic slide into corruption, despotism," *The Ottawa Citizen*, 13 December 1988, A9.

35. Hachten, *op. cit.*, 40.

36. Steve Geiman, Society of Professional Journalists (SPJ) Press Notes <SPJ-L@psuvm.psu.edu>, 25 December 1995.

37. George Lessard, Mass Media Arts, Training, Creation, from Canadian Association of Journalists (CAJ) listserv, via Media for Development in Democracy <devmedia@listserv.uoguelph.ca>, 12 July 1996.

38. Geiman, *op.cit.*

39. Lessard, *op. cit.*

40. *Loc. cit.*

41. *Loc. cit.*

42. *Loc. cit.*

43. Geiman, *op. cit.*

44. From George Ayittey, from africadev <africadev@egroups.com>, 10 July 2000.

45. Jennie Dey, *Women in food production and food security in Africa* (Rome: FAO, 1984), 6.

46. Hughes, *loc. cit.*

47. Howard W. French, New York Times Service, "Under a remote dictator, Zaire nears disintegration," *International Herald Tribune*, 12 June 1995, 2A.

48. For a description of how Africa's sidewalk trade in used and stolen goods functions, see Janet MacGaffey et al., *The real economy of Zaire: The contribution of smuggling and other unofficial activities to national wealth* (London: James Currey Ltd., 1994).

49. Hachten, *op.cit.*, 52.

50. Terry A. Olowu, "Reportage of Agricultural News in Nigerian Newspapers," *Journalism Quarterly*, 61(1), 1990, 200.

51. *Op. cit.*, 199.

52. *Loc. cit.*

53. Ochs, *op. cit.*, 48–49.

54. John Merrill, *op. cit.*, 215.

55. *Loc. cit.*

56. Paul Ansah, Cherif Fall, Bernard Chindji Kouleu, and Peter Mwaura, *Rural Journalism in Africa* (Paris: UNESCO, 1981), 3.

57. Ochs, *op. cit.*, 49.

58. Ansah et al., *op. cit.*, 30.

59. Ansah et al., *op. cit.*, 29.

60. Herman Daly, "Advice for a would-be reformer," *Ceres*, 27(6), 10–12.

61. United Nations Department of Public Information (DPI), *World Media Handbook* (New York: UNDPI, 1995).

62. Jaya Weera, Division of Communications, UNESCO, personal interview, 11 May 1995.

63. Jennifer Pitte, Developing Countries Farm Radio Network, personal interview, 26 July 1996.

64. Livai Matarirano, "The Farm Radio Network: Communications with a difference," *Ecoforum*, 18(3), 1995, 6.

65. Ochs, *op. cit.*, 51.
66. Hachten, *op. cit.*, 49.
67. John Merrill, *op. cit.*, 214.
68. Hachten, *op. cit.*, 50.
69. For information, contact Charles Okigbo, African Council for Communication Education (ACCE), P.O. Box 47495, Nairobi, Kenya.
70. African Council on Communication Education, *ACCE Directory 1988* (Nairobi: ACCE, 1988).
71. UNESCO, *Final Report: International Meeting of Regional Communication Training Institutions for Communication Development* (Paris: UNESCO, 1991).
72. UNESCO, *Final Report*, 5.
73. *Op. cit.*, 22.
74. Mwaura, *op. cit.*, 87.
75. Mwaura, *op. cit.*, 87–88.
76. Sam Phiri, Nordic-SADC Journalism Center, personal communication, 8 July 1996.
77. Francois Querre, *A Thousand and One Worlds: A Rural Radio Handbook* (Rome: United Nations Food and Agriculture Organization, 1992).
78. John Merrill, *op. cit.*, 253.

Chapter Nine

1. This analogy was suggested to me by Dr. Ernest Sternglass, of the University of Pittsburgh, during a 1980 interview on the Three Mile Island nuclear disaster, later published as an article in *Harrowsmith* magazine.
2. Anne Larson, "Educating non-ag publics," on International Communicators in Agriculture Network <ican_mg@ecn.purdue.edu>, 19 February 1996.

Selected Bibliography

Books

African Council on Communication Education. *ACCE Directory: A Directory of Communication Training Institutions in Africa.* Nairobi: ACCE, 1988.

Allen, R.E., ed., *The Concise Oxford Dictionary of Current English.* 8th ed. Oxford: Oxford University Press, 1992.

Association for Education in Journalism and Mass Communication (AEJMC), *Journalism and Mass Communication Directory 1995–1996.* Columbia, S.C.: AEJMC, 1996.

_____. *Journalism and Mass Communication Directory 1994–1995.* Columbia, S.C.: AEJMC, 1995.

Ansah, Paul, et al. *Rural Journalism in Africa (Reports and Papers On Mass Communication No. 88).* Paris: UNESCO, 1981.

Baker, C. Edwin. *Advertising and a Democratic Press.* Princeton N.J.: Princeton University Press, 1994.

Barbier, Edward B., et al. *Elephants, Economics and Ivory.* London: Earthscan Publications Ltd., 1990.

Bender, Barbara. *Farming in Prehistory: From Hunter-gatherer to Food-Producer.* London: John Baker, 1975.

Bene, Rose, et al. *Development Broadcasters' Manual.* Toronto and Kuala Lumpur: Ryerson International Development Centre and Asia-Pacific Institute for Broadcasting Development, 1992.

Berry, Wendell. *Farming: A Handbook.* New York: Harcourt Brace and Co., 1970.

_____. *The Unsettling of America: Culture and Agriculture.* San Francisco: Sierra Club Books, 1977.

Brady, Nyle C., ed. *Agriculture and the Quality of Our Environment.* Washington, D.C.: American Association for the Advancement of Science, 1967.

Broadcasting Yearbook 1976. New Providence, N.J.: R.R. Bowker, 1976.

Broadcasting & Cable Yearbook 1994. New Providence, N.J.: R.R. Bowker, 1994.

Conway, Gordon R., & Pretty, Jules N. *Unwelcome Harvest: Agriculture and Pollution.* London: Earthscan Publications Ltd., 1991.

Costello, Robert B., ed. *Random House Webster's College Dictionary.* 3d ed. New York: Random House, 1995.

Crosby, Alfred W. *Ecological Imperialism: The Biological Expansion of Europe, 900–1900.* Cambridge: Cambridge University Press, 1986.

Dey, Jennie. *Women in Food Production and Food Security in Africa.* Rome: FAO, 1984.

Dikshit, Kiranmani A., et al. *Rural Radio: Program Formats.* Paris: UNESCO, 1979.

Editor & Publisher International Yearbook, 1975. New York: Editor & Publisher Co., Inc., 1975.

Editor & Publisher International Yearbook, 1995. New York: Editor & Publisher Co., Inc., 1995.

Editor & Publisher International Yearbook, 1999. New York: Editor & Publisher Co., Inc., 1999.

Egginton, Joyce. *The Poisoning of Michigan.* New York: W.W. Norton & Co., 1980.

El-Ghonemy, Riad. *The Dynamics of Rural Poverty.* Rome, FAO: 1986.

Evans, James F., & Salcedo, Rodolfo N. *Communications in Agriculture: The American Farm Press.* Ames, IA: Iowa State University Press, 1974.

FAO. *Reorienting the Cooperative Structure in Selected Eastern European Countries, Vols. 1–7.* Rome: United Nations Food and Agriculture Organization, 1994.

_____. *Restructuring Agriculture in Eastern and Central Europe, Vols. 1–3.* Rome: United Nations Food and Agriculture Organization, 1994.

_____. *The State of Food and Agriculture* (SOFA94). Rome: United Nations Food and Agriculture Organization, 1994.

_____. *The State of Food and Agriculture* (SOFA95) Rome: United Nations Food and Agriculture Organization, 1995.

Fowler, H.W., & Fowler F.G., eds. *The Concise Oxford Dictionary of Current English,* 5th ed. Oxford: Oxford University Press, 1964.

Fox, Nicols. *Spoiled: The Dangerous Truth about a Food Chain Gone Haywire.* New York: BasicBooks, 1997.

Fukuoka, Masanobu. *The One-straw Revolution.* Emmaus, Penn.: Rodale Books Inc., 1978.

_____. *The Natural Way of Farming: The Theory and Practice of Green Philosophy* (2nd ed.) Tokyo: Japan Publications Inc., 1985.

George, Susan, & Sabelli, Fabrizio. *Faith and Credit: The World Bank's Secular Empire.* London: Penguin Books, 1994.

Giffard, C. Anthony. *International Press Coverage of the Food and Agriculture Organization.* Rome: United Nations Food and Agriculture Organization, 1992.

Hachten, William A. *The Growth of Media in the Third World: African Failures, Asian Successes.* Ames, Ia.: Iowa State University Press, 1993.

Hansson, Lennart; Fahrig, Lenore; and Merriam, Gray, eds. *Mosaic Landscapes and Ecological Processes.* London: Chapman & Hall, 1995.

Hardoy, Jorge; Mitlin, Diana; and Satterthwaite, David. *Environmental Problems in Third World Cities.* London: Earthscan Publications Ltd., 1993.

Harter, Eugene C. *Boilerplating America: The Hidden Newspaper.* Lanham, Md.: University Press of America, 1991.

Haugerud, Angelique. *The Culture of Politics in Modern Kenya.* Cambridge: Cambridge University Press, 1995.

Hippchen, Leonard J., & Yim, Yong S. *Terrorism, International Crime and Arms Control.* Springfield, Ill.: Charles C. Thomas, 1982.

Hughes, Anne. *The Diary of a Farmer's Wife: 1796–1797.* Ed. Jeanne Preston. London: Penguin Books, 1991.

Hughes, James, ed. *The Larousse Desk Reference.* New York: Larousse Kingfisher Chambers Inc., 1995.

Huxley, Elspeth. *White Man's Country.* London: Chatto & Windus, 1935.

Johnson, Lorraine. *The Ontario Naturalized Garden.* Vancouver, B.C.: Whitecap Books, 1995.

Kingston-Mann, Esther, & Mixter, Timothy, eds. *Peasant Economy, Culture and Politics of European Russia, 1800–1921.* Princeton, N.J.: Princeton University Press, 1991.

Krukones, James H. *To The People: The Russian Government and the Newspaper Sel'skii Vestnik (Village Herald), 1881–1917.* New York: Garland, 1987.

Lang, Tim, & Hines, Colin. *The New Protectionism.* London: Earthscan Publications Ltd., 1993.

Lawrence, Robert de T. *Rural Mimeo Newspapers.* Paris: UNESCO, 1963.

Lazer, William. *Handbook of Demographics For Marketing and Advertising.* New York: Lexington Books, 1994.

MacGaffey, Janet, et al. *The Real Economy of Zaire: The Contribution of Smuggling and Other Unofficial Activities to National Wealth.* London: James Currey Ltd., 1994.

Merrill, John C., ed. *Global Journalism: Survey of International Communication* (3d ed.) White Plains, N.Y.: Longmans Publishers, 1995.

Merrill, Richard, ed. *Radical Agriculture.* New York: New York University Press, 1976.

Murray, John. *The Russian Press From Brezhnev to Yeltsin: Behind the Paper Curtain*. Dublin, Ireland: Trinity College Press, 1994.

Mwaura, Peter. *Communication Policies in Kenya*. Paris: UNESCO, 1980.

Nelson, J.G. *Man's Impact on the Western Canadian Landscape*. Toronto: McClelland & Stewart Ltd., 1976.

O'Connor, Raymond J., & Watson, Donald. *Farming & Birds*. Cambridge: Cambridge University Press, 1986.

Ontario Ministry of Agriculture, Food, and Rural Affairs. *Kemptville College Calendar, 1995–1997*. Kemptville, Ont.: Kemptville College, 1995.

Querre, Francois. *A Thousand and One Worlds: A Rural Radio Handbook*. Rome: United Nations Food and Agriculture Organization, 1992.

Reeves, Geoffrey. *Communications and the Third World*. London: Routledge, 1993.

Repetto, R. *Paying the Price: Pesticide Subsidies in Developing Countries*. Washington: World Resources Institute, 1985.

Ricci, Nino. *Lives of the Saints*. Toronto: McKay, 1991.

Ryvkina, R., & Turovskiy, R. *The Refugee Crisis in Russia*. Toronto: York Lanes Press, 1993.

Shane, Scott. *Dismantling Utopia: How Information Ended the Soviet Union*. Chicago: Ivan R. Dee, 1994.

Statistical Service Unit, Policy Analysis Branch. *1993 Agricultural Statistics for Ontario*. Toronto: Ontario Ministry of Agriculture, Food and Rural Affairs. August 1994.

Steinbeck, John. *East of Eden*. London: Penguin Books Ltd., 1979.

Trzebinski, Errol. *The Kenya Pioneers*. London: William Heinemann Ltd., 1985.

UNESCO. *Developing Information Media in Africa: Press, Radio, Film, Television*. Paris: UNESCO, 1962.

_____. *Final Report: International Meeting of Regional Communication Training Institutions for Communication Development*. Paris: UNESCO, 1991.

United Nations Department of Public Information (DPI). *World Media Handbook*. New York: UNDPI, 1995.

Valbuena, Victor T., ed. *Mahaweli Community Radio: A Sri Lankan Experiment in Broadcasting and Development* Singapore: Asian Mass Communication Research and Information Centre, 1993.

Wallerstein, Emmanuel. *The Modern World System I: Capitalist Agriculture and the Origins of the European World Economy in the 16th Century*. San Diego, Calif.: Academic Press Inc., 1974.

Ward, William B. *Reporting Agriculture through Newspapers, Magazines, Radio, Television*. Ithaca, N.Y.: Comstock Publishing (Cornell University Press), 1959.

Williams, Raymond. *The Country and the City*. London: The Hogarth Press, 1993.

Wilson, Barry K. *Farming the System: How Politicians and Producers Shape Canadian Agricultural Policy.* Saskatoon, Saskatchewan: Western Producer Prairie Books, 1990.

Articles and Reports

Agocs, Peter, & Agocs, Sandor. "The change was but an unfulfilled promise": Agriculture and the rural population in post-communist Hungary. *East European Politics and Societies,* 8(1), 1994: 32–57.

African Rural and Urban Studies, "Special issue: Mass media and democratization in sub-Saharan Africa (part 1)," 4(1), 1997 (published 2000).

Agriculture and Agri-food Canada. *National Environment Strategy for Agriculture and Agri-food: A Report Prepared for Federal and Provincial Ministers of Agriculture.* Ottawa: Agriculture and Agri-food Canada, 1995.

————. "Future directions for Canadian agriculture and agri-food." Ottawa: September 29, 1994.

Akse, Gerard. *Rural Newspaper Development in Kenya.* The Hague: Graphic Media Development Centre, 1983.

Anderson, Clifton. "Farm issues and the media." *Editor & Publisher,* 121(49).

Appelgren B., & Burchi, S. "The Danube's blues." *Ceres,* 27(6).

Appleby, Timothy. "Rising stability puts Costa Rica at crossroads." *The Globe and Mail,* 6 July 1996.

Arus, Ben, & Warren, Marcus. "Media chief arrested in press purge by Kremlin." *The Electronic Telegraph,* 14 June 2000.

Baeza Lopez, Patricia. "A new plant disease: Uniformity." *Ceres,* 26(6), 1994: 41–47.

Balter, Michael. "Does anyone get GATT?" *Columbia Journalism Review,* May-June 1993.

Bianco, Miriam. "No more finger in the dike: Holland turns back history's tide." *Ceres,* 26(1), 1994: 12–13.

Blair, David. "Land grab in final phase." *The Daily Telegraph,* July 14, 2000.

Blauer, A.J. "Down the tubes: the dwindling future of private radio news." *Ryerson Review of Journalism,* Summer 1995.

Boafo, S.T. Kwame, & Arnaldo, Carlos A. The UNESCO communication program: Building capacity and protecting press freedom. *Media Development,* XLII(1), 1995: 3–7.

Bookchin, Murray. "Radical agriculture," in *Radical Agriculture.* Ed. Richard Merrill. New York: New York University Press, 1976: 3–13.

Bradbear, Nicola. "Protecting bees from pesticides." *Ceres,* 24(3), 1992: 47–48.

Braun, Armelle. "Sustaining the soil: Models for a workable future." *Ceres,* 22(2), 1990: 10–15.

Burds, Jeffrey. "The social control of peasant labor in Russia: The response of village communities to labor migration in the Central Industrial Region, 1861–1905." *Peasant Economy, Culture and Politics of European Russia, 1800–1921.* Eds. Esther Kingston-Mann and Timothy Mixter. Princeton, N.J.: Princeton University Press, 1991.

Canadian Press. "U.S. gears up for food fight with Canada." *The Ottawa Citizen,* 30 January 1996.

———. "Nunavut files court challenge to claim Inuit exemption from federal gun law." 22 June 2000.

Carter, Kim & McClellan, Scott. "Time is no friend to Poland's peasant farmers." *Ceres,* 23(2), 1991: 20–27.

Cave, Shane. "Rethinking a way of life." *Ceres,* 27(2), 38–40.

Civici, Adrian, Kercini, Donika & Griffin, Michael. "A nation of postage stamps." *Ceres,* 26(2), 1994: 40–42.

Cobb, Clifford, Halstead, Ted, & Rowe, Jonathan. "If the GDP is up, why is America down?" *The Atlantic Monthly,* October 1995.

Collinson, Mike. "Green evolution." *Ceres,* 27(4), 23–28.

Communications Branch, Agriculture and Agri-food Canada. *Product evaluation.* Ottawa: Agriculture and Agri-food Canada, Summer 1994.

Corbett, Julia B. "Rural and urban newspaper coverage of agriculture." *Journalism Quarterly,* 69(4), 1992: 929–37.

Cowan, Sharon L. "Ships on an unknown sea." *Ceres,* 26(3),1994: 18–23.

Craumer, Peter R. "Regional patterns of agricultural reform in Russia." *Post-Soviet Geography,* 35(6), 1994: 329–51.

Department of Agricultural Journalism, University of Wisconsin-Madison. Course prospectus, "Agricultural Journalism Department: Scientific, technical and agricultural communication." 1995.

Dunlop, John B. "Will the Russians return from the near-abroad?" *Post-Soviet Geography,* 35(4), 1994: 204–15.

El-Ghonemy, Riad, & Avis, Jeremy. "Food security=social security." *Ceres,* 27(2).

Elgie, Stewart. "Bite: endangered species need a law that gives them some." *Environment Views,* 18(1).

Elliot, Louise. "Crack on rise in rural Canada, experts warn: more young people using." *Canadian Press,* 23 July 2000.

Fagan, Drew. "Canada defends farm tariffs." *The Globe and Mail,* 30 April 1996.

Fitzgerald, Mark. "A year of turmoil." *Editor and Publisher.* 129(1).

French, Howard W. New York Times Service. "Under a remote dictator, Zaire nears disintegration." *International Herald Tribune,* 12 June 1995, 2A.

Gettings, T.L., & DeVault, George. "The Russian revelation." *The New Farm,* 17(4).

Gillman, Helen, & Grimaux, Helen. "Utopia as judgement: The agricultural visionary's dream is also a criticism—one we can't afford to ignore." *Ceres*, 24(6), 1992: 15–23.

Goldburg, Rebecca. "Pause at the amber light." *Ceres*, 27(3), 1995: 21–25.

Griswold, William F., & Swenson, Jill D. "Development news in rural Georgia newspapers: A comparison with media in developing nations." *Journalism Quarterly*, 69(3), 1992: 580–90.

Griffin, Michael. "The seeds of death." *Ceres*, 26(1), 1994: 36–41.

Harrington, Carol. Canadian Press, "Calgary cancels cowboy festival due to lack of interest." *Sympatico News Express*, 4 July 2000.

Harrowsmith, VII(2). "Guardians of a way of life: a fresh perspective on marketing boards."

Hays, Robert G., & Reisner, Ann. "Feeling the heat from advertisers: Farm magazine writers and ethical pressures." *Journalism Quarterly*, 67(4), 1990: 936–42.

_____. "Farm journalists and advertiser influence: Pressures on ethical standards." *Journalism Quarterly*, 68(1/2), 1991: 172–78.

Herbert, John. "The narrowing of the options: The changing face of the seed trade." *Ceres*, 26(6), 1994: 42–45.

Ibbitson, John. "The high cost of common sense." *The Globe and Mail*, 24 May 2000.

ILEIA, Center for Research and Information Exchange in Ecologically Sound Agriculture. *Working for ecologically sound agriculture*. Leusden: the Netherlands Ministry of Development Cooperation, 1996.

International Labor Organization (ILO). "Trouble in the transition." *World of Work*, No. 12 (1995).

James, Natalie. "Report arouses concern over government role in E. coli deaths." *Canadian Press*, 28 July 2000.

Kotz, Nick. "Agribusiness." *Radical agriculture*, ed. Richard Merrill. New York: New York University Press, 1976.

Lajoie, Don. "Proposed greenhouse big as four farms." *The Windsor Star*, 23 June 2000.

Lang, Tim, & Hines, Colin. "A disaster for the environment, rural economies, food quality and food security." *Ceres*, 27(1).

Martin, R.B. "A voice in the wilderness." *Ceres*, 26(6).

Matarirano, Livai. "The Farm Radio Network: Communication with a difference." *Ecoforum*, 18(3), 1995: 6–7.

McMillan, Robert T. "Some observations on Oklahoma population movements since 1930." *Rural Sociology*, 1(3), 1935: 332–43.

Miljan, Lydia. "Agriculture: Television coverage of the farm crisis." *On Balance*, 7 (July-August), 1992: 1–8.

Mollison, Bill, quoted in John Madeley. "A design science with an ethic." *Ceres*, 24(6).

Nagle, Patrick. Southam News, "Kenya: Tragic slide into corruption, despotism." *The Ottawa Citizen*, 13 December 1988.

National Research Council, Committee on the Role of Alternative Farming Methods in Modern Production Agriculture, Board on Agriculture. *Alternative Agriculture*. Washington, D.C.: National Academy Press, 1989.

Olowu, Terry A. "Reportage of agricultural news in Nigerian newspapers." *Journalism Quarterly*, 67(1), 1990: 195–200.

Patnaik, Ajay. "Agriculture and rural out-migration in Central Asia, 1960–1991." *Europe-Asia Studies*, 47(1), 1995: 147–69.

Pawlick, Thomas. "The cause and its effects." *Harrowsmith*, VII(2), 1982: 26–36.

_____. "The essential ploughman." *Harrowsmith*, IV(7), 42–49, 1980: 104–106.

Petrikov, A. "Rural society and agrarian reform." *Problems of Economic Transition*, 37(6), 1994: 19–28.

Pitman, Dick. "Wildlife as a crop." *Ceres*, 22(1).

Polk, Peggy. "'Laissez-faire' without a net." *Ceres*, 26(3), 1994: 32–35.

Postel, Sandra. "The waters of strife: The concept of fresh water as a free good is fast becoming an economic and political anachronism." *Ceres*, 27(6), 1995: 19–24.

Pugh, Terry. "Thousands of family farmers will become casualties to this adjustment." *Ceres*, 27(1), 1995: 28–33.

Reisner, Ann. "An overview of agricultural communications programs and curricula." *Journal of Applied Communications*, 74(1), 1990: 8–17.

_____. "Course work offered in agricultural communication programs." *Journal of Applied Communications*, 74(1), 1990: 18–25.

Reisner, Ann, & Walter, Gerry. "Agricultural journalists' assessments of print coverage of agricultural news." *Rural Sociology*, 59(3), 1994: 525–37.

Reuters Limited. "Police search for Russian cameraman." 11 July 2000.

Richardson, Don (1995). Community electronic networks: Sharing lessons learned in Canada with our African colleagues. Department of Rural Extension Studies, University of Guelph, Guelph, Ontario: paper published on World Wide Web URL <http://htdg.uoguelph.ca~res/occasional_papers/occasional_papers.html>.

Scanlon, T. Joseph. "A study of the contents of 30 Canadian daily newspapers for Special Senate Committee on Mass Media." Carleton University, 31 October 1969.

Shiva, Vandana. "Mistaken miracles." *Ceres*, 27(4).

Spears, Tom. "Swill from the swine." *The Windsor Star*, 10 June 2000.

Steffen, Alex, & Atkisson, Alan. "The Netherlands' radical, practical Green Plan." *Whole Earth Review*, 87, 1995: 94–99.

Stroud, Polly. "Africa's wave of the future, or a backwash from the past?" *Ceres*, 27(4).

Switzer-Howse, K.D., & Coote, D.R. "Agricultural practices and environmental conservation." Ottawa: Agriculture Canada, 1984.

Taiganides, E. Paul. "The animal waste disposal problem." *Agriculture and the quality of our environment*. Ed. Nyle C. Brady. Washington, D.C.: American Association for the Advancement of Science, 1967.

Tangermann, Stefan. "A major step in a good direction." *Ceres*, 27(1).

Temple-Raston, Dina. "Family farms can't beat corporate muck." *USA Today*, 23 May 2000.

Toth, Andras. "The social impact of restructuring in rural areas of Hungary: disruption of security or the end of the rural socialist middle class society?" *Soviet Studies*, 44(6), 1992: 1039–43.

Trickey, Mike. "Reforms sow seeds of despair on failing farms." *Ottawa Citizen*, November 21, 1995: D-12.

Turk, Linda. "No sacred cows in the NAFTA era." *The Globe and Mail*, 25 April 1996.

Turning Point Project. "Unlabelled, untested and you're eating it." Advertisement No. 2 in a series, 14 October 1999.

Van Tighem, Kevin. "Save the gopher." *Environment Views*, 18(1), 1995: 16–19.

Wadhams, Nick. "Russia said restricting media." *The Associated Press*, 1 August 2000.

Walston, James. "C.O.D.E.X. spells controversy." *Ceres*, 24(4).

Webb, John. Krasnodar: A case study of the rural factor in Russian politics. *Journal of Contemporary History*, 29, 1994: 229–60.

Wegren, Stephen K. "Dilemmas of agrarian reform in the Soviet Union." *Soviet Studies*, 44(1), 1992: 3–36.

_____. "Regional development of Russian private farms: a comment." *Post-Soviet Geography*, 35(3), 1995: 176–84.

_____. "Rural migration and agrarian reform in Russia: A research note." *Europe-Asia Studies*, 47(5), 1995: 877–88.

_____. "Rural reform and political culture in Russia." *Europe-Asia Studies*, 46(2), 1994: 215–41.

Weiss, Rick. "Insect Bambi threatened by gene-altered corn." *The Ottawa Citizen*, 20 May 1999.

Wilson, Tamar Diana. "What determines where transnational labor migrants go? Modifications in migration theories." *Human Organization*, 53(3), 1994: 269–76.

Wylynko, David. "The rate debate: Will the end of the transportation subsidy for prairie wheat lead to more sustainable farming practices on the Great Plains?" *Nature Canada* 25(1).

Personal Communication

Alda, Wayne F. <wfalda@aol.com>. Personal e-mail, 25 March 1996.

Armstrong, Max. <maxarm@aol.com>. WGN, U.S. Farm Report, Chicago. Personal e-mail, 31 October 1995.

Aycock, Frank A., Associate Professor of Journalism, American University in Bulgaria. Personal correspondence. 30 April 1995.

Bierhoff, Jan. <bierhoff@ejc.nl>. Managing director, European Journalism Center. Personal e-mail, 12 July 1995.

Billings, Linda. <billings@anamorphosis.usra.edu>. Personal e-mail, 25 September 1995.

Bridges, Professor Lamar W. Department of Journalism and Printing, East Texas State University. Personal communication. 22 January 1996.

Brodell, Jay. <brodellj@mscd.edu>. Personal e-mail, 24 April 1995.

Cavanagh, Kevin. Personal interview. 28 June 1996.

Cooke, Michael. Personal interview. 28 June 1996.

da Costa, Peter. <ipsdc@gn.apc.org>. Personal e-mail, 2 May 1995.

Davis, Bob. <ift@soli.inav.net>. Personal e-mail, 30 November 1995.

Hendry, Peter. <phendry@uoguelph.ca>. Personal e-mail, 19 October 1995.

Kappel, Tana. <apbtk@msu.oscs.montana.edu>. Personal e-mail, 25 March 1996.

Milito, Ron. Agriculture and Agri-Food Canada. Personal correspondence. 18 October 1995.

Neil, John. <crs0148@inforamp.net>. Personal e-mail, 15 March 1996.

Owen, Hugh. <owen@ornet.or.uoguelph.ca>. Personal e-mail, 28 October 1995.

Phiri, Sam. Nordic-SADC Journalism Center. Personal communication. 8 July 1996.

Pilugin, Nicholas. <nwpilugin@glas.apc.org>. USIA Media Assistance Clearinghouse. Personal e-mail, 6 February 1996.

Pitte, Jennifer. Developing Countries Farm Radio Network. Personal interview. 26 July 1996.

Queck, Paul. Personal correspondence. 10 November 1995.

Rivoire, Brigid. Agriculture and Agri-Food Canada. Personal notes from panel discussion "Agri-Food: the forgotten beat." Ottawa, 29 June 1993.

Roberts, Owen. <owen@ornet.or.uoguelph.ca>. Personal e-mail. 18 September 1995.

Romahn, James. <jromahn@web.apc.org>. Personal e-mail, 15 March 1996.

Simkova, Viera. <trend@savba.savba.sk>. Personal e-mail, 4 October 1995.

Strathdee, Michael. <strathdeem@kwrmsnt.cmail.southam.ca>. Personal e-mail, 14 March 1996.

Toros, Hilmi, Chief, Press Section. United Nations Food and Agriculture Organization (FAO). Personal interview. 17 June 1996.

Weera, Jaya. Division of Communications, UNESCO. Personal interview. 11 May 1995.

List Servers

Agricultural Communicators in Education (ACE) <ace@gnv.ifas.ufl.edu> reported on FSUMedia <fsumedia@sovam.com>. 8 December 1995.

Ayittey, George from africadev <africadev@egroups.com>. 10 July 2000.

Brown, Betty P., ed. <fsumedia@sovam.com>. "Russian ag communicator update." 1 December 1995.

Center for Foreign Journalists: CFJ Clearinghouse, reported on East European Media List <eemedia@mcfeeley.cc.utexas.edu>. 14 July 1995.

Editor. <editor@uu6.psi.com>. 22 November 1995.

Geiman, Steve. Society of Professional Journalists (SPJ) Press Notes <SPJ-L @psuvm.psu.edu>. 25 December 1995.

Guiragos Manoyan. <fsumedia@sovam.com>. 30 September 1995.

International Freedom of Expression Exchange Clearinghouse <fsumedia @sovam.com>. 11 July 1995.

International Freedom of Expression Exchange Clearinghouse <fsumedia @sovam.com>. 17 October 1995.

Johnson, Eric. Internews <fsumedia@sovam.com>. 6 July 1995.

Lessard, George. Mass Media Arts, Training, Creation. Canadian Association of Journalists (CAJ) listserv, via Media for Development in Democracy <devmedia@listserv.uoguelph.ca>. 12 July 1996.

Moscow *Times*. <fsumedia@sovam.com>. 15 October 1995.

OMRI Daily Digest. <belin@omri.cz> via <fsumedia@sovam.com>. 18 July 1995.

OMRI This Week. <fsumedia@sovam.com>. 2 December 1995.

OMRI Daily Digest, reported by Laurie Belin. <belin@omri.cz>. on FSUMedia <fsumedia@sovam.com>. 6 July 1995.

OMRI Daily Digest <fsumedia@sovam.com>. 11 July 1995.

OMRI Daily Digest <fsumedia@sovam.com>. 24 November 1995.

Pilugin, Nicholas. USIA Media Assistance Clearinghouse <nwpilugin@glas. apc.org> reported on FSUMedia <fsumedia@sovam.com>. 17 July 1995.

Rokityanski, Yuri. "TV should be produced not only in metropolis." International Media Centre/Internews <rockit@sovam.com>. FSUMedia <fsumedia@sovam.com.> 6 July 1995.

Rural Advancement Foundation International <http://www.rafi.org> 21 July 2000.

Rural Advancement Foundation International <http://www.rafi.org> 28 July 2000.

RUSAG-L: Current Events #79 <RUSAG-L@UMDD.UMD.EDU>. 8 July 1996.

RUSAG-L: Current Events #73 <RUSAG-L@UMDD.UMD.EDU>. 15 February 1996.

USIA Media Assistance Clearinghouse, reported on FSUMedia List <fsumedia @sovam.com>. 5 July 1995.

Index

Abacha, Sani, 167
Advertising, influence on media,
    94–98
Africa
    farm journalism, 131–58
    journalism training, 155–58
    rural population loss, 147–53
    since collapse of communism,
        132–36
    state of journalism in, 136–37
    threats to journalism, 143–47
African Council on
        Communication Education, 6,
        155
African Farmer, 152
African game areas, 55–57
Afrique Agriculture, 152
Agent Orange herbicide, 46
Agriculture
    in Africa, 132–36
    call for radical changes, 78–80
    clash of practices, 75–79
    as cultural endeavor, 11–15
    current state in Russia, 115–22
    decline in trade, 37
    as declining industry, 36–39
    economics and trade issues,
        23–39

environmental issues, 41–73
false Soviet reports, 112–15
industrialization of, 17–22,
    75–79
lack of media coverage, 83–90
lack of media training, 6–7
oligopoly situation, 29–32
radical decline in farm
    population, 15–22
shrinkage of employment, 93–94
under siege, 15–22
as source of water pollution, 78
sustainable, 75–79
under-reporting, 2–5
use of pesticides, 45–40
Agriculture and Agri-Food Canada
    Communication Branch
    Product Evaluation, 86
Agriculture Canada, 48, 58, 60
Agri-Forum, 152
Agri-information, 152
Agrochemicals, 45–50
Albania, 116
Alda, Wayne F., 87–88
American Agricultural Editors'
    Association, 3
American Association for the
    Advancement of Science, 48

American literature, 12–13
Amin, Idi, 135, 143
Anderson, Clifton, 85
Angola, 145
Animal waste, 48–50
Aral Sea basin, 61
Arap Moi, Daniel, 131–32, 135, 137, 143, 167
Archer Daniels Midland, 30
Armstrong, Max, 83, 87
Artists, 13
Asgrow Seed Company, 67, 70
Association for Education in Journalism and Mass Communication, 6, 102, 128–29
Aswan High Dam, Egypt, 61
Atkins,George, 153

Baker, C. Edwin, 94–95, 97
Balter, Michael, 88–89
Bangladesh, 62
Bender, Barbara, 11
Berry, Wendell, 19–20, 22
Biodiversity, loss of, 62–68
Blair and Ketchum's Country Journal, 99
Blankenberg, William, 98
Blauer, A. J., 92
Bokassa, Emperor, 135
Borlaug, Norman, 76
Broadcasting and Cable Yearbook, 3
Broadcasting Yearbook, 3
Brodell, Jay, 130
Bookchin, Murray, 15, 18, 19, 63
Bulgarian Agricultural Academy, 122
Bunge and Borne, 30
Burds, Jeffrey, 107
Bureau of Labor Statistics, 93
Burkina Faso, 156
Burundi, 146
Butz, Earl, 21, 33

Cameroon, 146
Campfire program, 55–56
Canada
  drop in rural population, 16
  gun control and Inuit, 101–102
  North American Free Trade Agreement case against, 32–34
  perception of agricultural decline, 83–103
Canadian National Farmers Union, 27
Canadian Press, 50
Canadian Senate Special Committee on Mass Media, 89
Cargill Grain, 28–29, 30, 66
Cattle, contaminated, 1–2
Cattle feed, 1–2
Cavanagh, Kevin, 91
Ceres magazine, 78–79, 152
Cervantes, Miguel de, 12
Chavez, Cesar, 108
Chicago School of Economics, 27
Chopra, Shiv, 77–78
Classical economics, 27–28
Coca-Cola Company, 95
Codex, Alimentarius, 72
Collectivization of agriculture, 110
Commonwealth of Independent States
  crime and instability, 123–28
  current state of, 115–22
  journalism education, 128–30
Communal Areas Management Program for Indigenous Resources, Africa, 55–56
Communism, collapse of, 132–36
Consultative Group for International Agricultural Research, 76
Consumer Price Index, 94
Contaminated-water deaths, 39, 50
Continental, 30

Conway, Gordon, 43–36
Cooke, Michael, 92, 93
Cost-price squeeze, 29–32
Côte d'Ivoire, 145
Country and the City, The
   (Williams), 13
Crosby, Alfred W., 68–69

Da Costa, Peter, 7
Daily Nation, Kenya, 142
DANIDA, 160
Davis, Bob, 89
Decree on Land of 1918, Russia,
   114
Des Moines *Register,* 97
Detroit *Free Press,* 2
Developing countries, since
   collapse of communism,
   132–36
Developing Countries Farm Radio
   Network, 153
Dey, Jennie, 147
Dickens, Charles, 35
Diouf, Jacques, 152
Duke University, 130
Dunlop, John B., 117–18
Du Pont, 95
Dutch Institute for Low External
   Input Agriculture, 78–79
Dynamics of Rural Poverty (El-
   Ghonemy), 35–36

Eastern Canada Farm Writers
   Association, 86
Eastern Europe, 6
   journalism education, 128–30
East of Eden (Steinbeck), 13
E-coli outbreak, 50
Editor & Publisher International
   Yearbook, 3, 89–90
Egginton, Joyce, 1–2
El-Ghonemy, Riad, 35–36, 37,
   38

English literature, 12–13
Environmental issues, 41–73
   exotic species, 68–73
   loss of biodiversity, 62–68
   pollution, 45–50
   soil depredation, 58–60
   use of pesticides, 45–40
   waste of freshwater resources,
      60–62
   wildlife habitat destruction,
      51–58
European Journalism Center,
   129–30
European Union, drop in rural
   population, 16
Exotic species, 68–73

Family farm
   loss of, 17
   social catastrophe, 34–36
Farm Bureau Services, 1–2
Farm incomes, 94
Farming and Birds (O'Connor &
   Shrubb), 51
Farming and Prehistory (Bender),
   11
Farm journalism
   in Africa, 131–58
   inadequacy of, 83–90
   reasons for decline in coverage,
      90–93
   training in, 102–103
   UNESCO report on, 150–52
Feruzzi, 30
Figes, Orlando, 110
Food and Agriculture Organization,
   16, 24, 37, 72, 152
   report on African farming,
      134–35
   on trade rules, 25–26
Food safety rules, 31
Food safety standards, 71–73
Free trade principles, 27–29

Freshwater resources, 60–62
Friedman, Milton, 27
Fukuoka, Masanobu, 11, 41, 76

Gale Directory of Publications and Broadcast Media, 86
Ganges River, 62
Gene-altered organisms, 70–71
General Agreement of Tariffs and Trade, 16–17, 23–24, 30
agriculture agreement, 24
and food safety standards, 71–73
free trade principles, 27–29
and marketing boards, 31–32
reform *versus* liberalization, 25–26
Genetically engineered products, 73
Genetically engineered seeds, 67
Georgics (Virgil), 12
Giono, Jean, 12
Global debt crises, 23
Global Journalism (Merrill), 4
Goldburg, Rebecca, 68, 70
Gorbachev, Mikhail, 114–15
Green Revolution, 44, 65, 76
Gresev, Ivan Petrovich, 112
Gridasov, Ivan, 121
Gross Domestic Product
as economic yardstick, 38
fall in Russia, 120
Growth hormones, 72
Gulf War, 95–96
Gun control legislation, Canada, 101–102
Gusinsky, Vladimir, 125

Haidulsky, Pavel, 122
Hard Times (Dickens), 35
Harris, Mike, 39
Harrowsmith Magazine, 99
Hatchen, William A., 140–41, 143, 148, 154, 155

Hayden, Margaret, 77
Hendry, Peter, 85–86, 92, 98
Herbicides, 46
Hines, Colin, 27, 35
Hoening, Thomas, 34
Hume, David, 27

ICI, 66
Illiteracy, in Africa, 141, 147–53
India, 62
Industrial agriculture, 75–79
Industrialized farming, 17–22
Insecticides, 46
Institute for Low External Input Agriculture, Netherlands, 78–79
International Center for Maize and Wheat Improvement, 44
International Commission on Radio-Television Policy, 130
International Council for Research in Agroforestry, 157
International Development Research Center, 153
International Meeting of Regional Communication Training Institutions for Communication Development, 155
International Monetary Fund, 23, 27, 116
International Rice Research Institute, 44
Inter Press Service, 6
Interventionist economics, 28
Inuit culture, 102
Invisible farm, 708
Iowa, decline in number of newspapers, 91
Iowa Farm Today, 89

Japan, drop in rural population, 16

Journalism
see also Farm journalism
in Africa, 136–37
failure to cover farming, 708
lack of training in agriculture,
6–7
threats to, in Africa, 143–47
Journalism education/training
in Africa, 155–58
in Russia and Eastern Europe,
128–29
training in agriculture, 102–103

Kappel, Tone, 88
Kark, Tom, 92
Kemptville College of Agriculture, 6
Kenya, 16, 131–32, 135, 145–46
Kenya Farmer, 152
Kenya Institute of Mass
Communication, 167
Kenyatta, Jomo, 137, 143
Keynes, John Maynard, 28
Khruschev, Nikita, 112, 115
Kingston-Mann, Esther, 105
Kniazhytsky, Mykola, 126–27
Kotz, Nick, 17, 18, 19
Kuchma, Leonid, 122
Kulaks, 108

Laissez-faire approach, 27–28
Lang, Timothy, 27, 35
Languages of Africa, 141
Lapointe, Kirk, 86
Latin America, since collapse of
communism, 132–36
Lazarenko, Pavlo, 122
Lazer, William, 93–94
Leopold Center for Sustainable
Agriculture, 160
Less-developed countries, since
collapse of communism,
132–36

Liberia, 143–44
Limagrain, Inc., 66
Lives of the Saints (Ricci), 13
Louis Dreyfus, 30
Loyalist College of Applied Arts
and Technology, 6

MacMillan, Whitney, 28, 30
Macy, Janet, 128
Madeira, 68–79
Magazines
with farm reporting, 99–100
influence of advertising, 94–98
Marketing boards
Canadian, 29–30
and General Agreement of
Tariffs and Trade, 31–32
U.S. view of, 34
Martin, R. B., 57
McMillan, Robert T., 120
McNamara, Robert S., 21
Media
current climate for, 98–102
influence of advertising, 94–99
lack of farm coverage, 83–90
lack of training in agriculture,
6–7
marketability problem, 99–102
urban-rural blind spot, 2–5
Merrill, John C., 4, 133, 134–35,
136, 138–39, 150, 154–55,
158
Michigan Chemical Corporation,
1–2
Michigan Farmer, 2
Mill, James, 27
Mill, John Stuart, 27
Miller, Gord, 50
Mitsui, 30
Molleson, Bill, 75
Monthly News, Kenya, 142
Mother Earth News, 99
Mugabe, Robert, 135

Murray, John, 112, 114–15
Mwaura, Peter, 137–38, 139–40
Myth, 14

Nagle, Patrick, 142–43
Nairobi Law Monthly, 142
National Association of Agricultural
   Journalists, 3
Netherlands plan, 78–80
New Protectionism (Lang and
   Hines), 27
Newspapers
   cost-cutting, 92–93
   decline in number of, 90–91
   demographic profile, 96–97
   influence of advertising, 94–98
   tabloid format, 92
New Zealand plan, 79–80
Nickerson, 66
Nigeria, 144–45
Nordic-SADC Journalism Center,
   160
North America
   drop in rural population, 16
   perception of agricultural
      decline, 83–103
North American Free Trade
   Agreement, 23–24, 29, 31
   farm case against Canada, 32–34
Nunavut, 101–102
Nutrimaster, 1–2

Ochs, Martin, 141–42, 149–50,
   151, 152
Okalik, Paul, 102
Oligopoly situation, 29–32
Oliver Twist (Dickens), 35
Open Media Research Institute,
   126
Owen, Hugh, 98–99

Painters, 13
Panoscope, 152

PBBs, 1–2
Permaculture Institute, 75
Pesticides, 45–50
Petroseed, 67
Piers Plowman (Langland), 12
Pilugin, Nicholas, 125–26, 127
Pioneer Hi-Bred, 66
Plant breeding, 63–65
Plant species, 63
Plekhanov, G. V., 109
Poisoning of Michigan (Egginton),
   1–2
Pollard, Bonnie, 2
Pollution, 45–50
Porto Santo, 68–79
Postel, Sandra, 61, 62
Potok, Ana, 119
Premium Standard, 34
Pretty, Jules, 43–36
Privatization in Russia, 116
Project Censored, 4, 87
Project Censored Canada, 87
Protectionism, 27–29
Pugh, Terry, 27, 28, 30, 31
Putin, Vladimir, 124, 125

Queck, Paul, 85

"Radical Agriculture" (Bookchin),
   15
Radio
   in Africa, 153
   influence of advertising, 94–98
Radio stations, 92
Reagan, Ronald W., 27–28, 36
Reisner, Ann, 100, 102–103
Reporting Agriculture (Ward),
   100
Ricardo, David, 27, 28
Ricci, Nino, 13
Rodale, J. I., 76, 128
Rodale, Robert, 76, 128
Rodale Institute, 128

Romahn, James, 83–85, 98
Royal Sluis, 67
Rural Advancement Foundation,
    67
Rural Development Ministry,
    Chad, 153
Rural economy, 23–339
    cost-price squeeze, 29–32
    decline of family farm, 34–36
    declining industry, 36–39
    micro-level example, 32–34
    new regime, 24–26
    revival of protectionism, 27–29
    under siege, 15–22
    trade negotiations and, 24–26
Rural Journalism in Africa
    (UNESCO), 5
Russia, 6
    agriculture, 105–30
    crime and instability, 123–28
    current state of farming, 115–22
    fall in Gross Domestic Product,
        120
    fall in rural population, 16
    journalism education, 128–30
    serfs, 106–109
Russian Agricultural Listserv,
    120
Ryvinka, R., 118

Sandoz, 66
Sasakawa Foundation, 66
Savia, 67
Seed trade, 66–67
Seminis seed corporation, 67
Serfs, 106–109
Shakespeare, William, 12
Shane, Scott, 110
Shanks, Bob, 95
Shiva, Vandana, 65, 76–77
Simkova, Viera, 6, 127
Slovak Club of Agricultural
    Journalists, 6

Smith, Adam, 27
Soil depredation, 58–60
South African Development
    Community, 147, 160
Soviet Union
    agriculture, 105–30, 110–11
    coverage of agriculture, 4–5
    false reports on farming, 112–15
    fall in rural population, 16
    under Gorbachev, 114–15
    victims of Stalin, 111–12
Stalin, Josef, 110–11
Steinbeck, John, 1313
Sub-Saharan Africa, 5
Sustainable agriculture, 75–79

Taifa Leo newspapers, 142
Tangermann, Stefan, 26
*Tanzania Agriculture and
    Livestock*, 156
Tanzania Information Services,
    156
Television, influence of advertising,
    94–98
Tenneco, Inc., 17
Terminator crops, 67
Thatcher, Margaret, 27–28
Thiongo, Ngugi wa, 12
Thoreau, Henry David, 16
Times of London, 96–97
Tolkien, J. R. R., 12
Toronto *Globe and Mail*, 32–34, 39
Toros, Hilmi, 127
Trade barriers, reduction in, 24–26
Transgenic organisms, 70–71
Trickle-down economics, 36
Turning Point Project, 73
Turovsky, R., 118

Uganda, 143
UNESCO, 6
    report on farm journalism,
        150–52

Union Farmer, 27
United States, perception of
    agricultural decline, 83–103
United States Agency for
    International Development,
    129
United States Department of
    Agriculture, 67, 76
United States Environmental
    Protection Agency, 78
United States Food and Drug
    Administration, 73
United States Information Agency,
    130
    Media Assistance Clearinghouse,
    125–26
United States National Association
    of Agricultural Journalists,
    87–88
University of Guelph, 102–103
University of Illinois Office of
    Agricultural Communication
    and Education, 3
Unwelcome Harvest (Conway &
    Pretty), 47
Upjohn, 66
Urbanization, 13
Urban-rural media blind spot, 2–5
Urban squalor, 35

Uruguay Round, 16–17, 23
USA Today, 34

Virgil, 12
Vodka protest of 1859, Russia, 109
Voice of Kenya, 131

Walkerton, Ontario, 39, 50
Wallerstein, Immanuel, 35
Wall Street Journal, 38
Walter, Gerry, 100
Ward, William B., 100
Water pollution, 78
Weera, Jaya, 153
Wegren, Stephen, 118
West African Riziculture
    Development Association, 157
Western Producer, 86
Wild life habitat destruction, 51–58
Williams, Raymond, 13, 35
Wilson, Barry, 86
Windsor Star, 20
World Bank, 23, 27
World Health Organization, 72

Zambia, 144
Zambia Institute of Mass
    Communications, 156
Zimbabwe, 55–57, 13

About the Author

Thomas F. Pawlick is an award-winning science journalist and teacher with more than 30 years experience in publishing and 10 in the classroom. A three-time winner of the Canadian Science Writers Association National Journalism Award, he has also received the National Magazine Award for agricultural reporting. He served for six years as Chief Editor of the United Nations Food and Agriculture Organization (FAO) flagship publication, *Ceres* magazine, during which time he traveled to more than 20 countries, observing farming methods and problems. In 1996 he earned his masters degree in agricultural journalism from Carleton University, Ottawa. He has owned and operated his own farms in Quebec and Ontario. Currently, he is an assistant professor of journalism at the University of Detroit.